Postharvest Handling and Technology

Postharvest Handling and Technology

Aliyah Wheatley

Larsen & Keller
www.larsen-keller.com

Postharvest Handling and Technology
Aliyah Wheatley
ISBN: 978-1-64172-126-4 (Hardback)

Larsen & Keller
Published by Larsen and Keller Education,
5 Penn Plaza,
19th Floor,
New York, NY 10001, USA

Cataloging-in-Publication Data

Postharvest handling and technology / Aliyah Wheatley.
p. cm.
Includes bibliographical references and index.
ISBN 978-1-64172-126-4
1. Farm produce--Handling. 2. Crops--Postharvest technology. 3. Farm produce--Storage. I. Wheatley, Aliyah.
S571 .P67 2019
630--dc23

Printed and bound in China.

For more information regarding Larsen and Keller Education and its products, please visit the publisher's website www.larsen-keller.com

Table of Contents

Preface

Postharvest handling is an important stage in agriculture. It is the stage of crop production that follows the stage of harvesting and includes the processes of cooling, sorting, cleaning and packing. A crop starts to deteriorate once it is separated from its parent plant. To maintain its quality for consumption or sale, an effective postharvest handling regime needs to be designed. This involves preventing moisture loss, slowing down chemical changes, keeping the product cool and delaying spoilage. Sanitation is also an important aspect of this process. This textbook provides comprehensive insights into the field of postharvest handling and technology. It elucidates new techniques and applications of postharvest handling and technology in a multidisciplinary approach. Those in search of information to further their knowledge will be greatly assisted by this textbook.

To facilitate a deeper understanding of the contents of this book a short introduction of every chapter is written below:

Chapter 1, In agriculture, the next stage of crop production after harvesting is postharvest handling, which involves the processes of cleaning, packing and cooling. The technology used in postharvest handling determines the final quality of the crops. This is an introductory chapter, which will analyze briefly all the significant aspects of postharvest technology. It includes the different processes involved in the postharvest handling of crops, such as food grading, culling, curing of roots, etc. **Chapter 2**, Post-harvest losses occur between the stages of harvest and human consumption. Losses of horticultural produce and grains are significant in the post-harvest sector. This chapter has been carefully written to provide an easy understanding of the varied aspects of postharvest losses in grains, fruits and vegetables. It includes vital topics such as vegetable pathogens, senescence in fruits and fruit pathogens. **Chapter 3**, This chapter closely examines the key processes and techniques of postharvest storage to provide an extensive understanding of agricultural postharvest technology. It includes various topics related to cold storage, frosting, freezing and chilling injury, temperature control with respect to different grains, fruits and vegetables and techniques of minimizing physical damage, besides others. **Chapter 4**, Science and technology have undergone rapid developments in the past decade, which has resulted in improved techniques of postharvest packaging in the field of agriculture. This chapter extensively discusses about the fundamentals of food packaging, use of sanitizers and disinfectants, removal of microorganisms, artificial waxing, preservation, vacuum packaging, food fortification, etc. **Chapter 5**, The most important goal of post-harvest handling is preservation of the agricultural produce by keeping the product cool, avoiding moisture loss and delaying any unwanted chemical change. Sanitation is another consideration. The diverse machinery that helps to ensure such objectives have been extensively covered in this chapter. These include cartooning machines, form, fill and seal machines, grain dryer, rice huller, hay rake, etc.

I would like to share the credit of this book with my editorial team who worked tirelessly on this book. I owe the completion of this book to the never-ending support of my family, who supported me throughout the project.

Aliyah Wheatley

1

Understanding Postharvest Technology

In agriculture, the next stage of crop production after harvesting is postharvest handling, which involves the processes of cleaning, packing and cooling. The technology used in postharvest handling determines the final quality of the crops. This is an introductory chapter, which will analyze briefly all the significant aspects of postharvest technology. It includes the different processes involved in the postharvest handling of crops, such as food grading, culling, curing of roots, etc.

Postharvest Handling of Crops

Postharvest handling integrates activities include harvesting, in-field handling, packinghouse operation, storage, transportation and ripening. Harvesting time is the earlier stage in postharvest handling practices. Due to food safety and quality concern, harvesting activities should be done during the coolest time of the day. Normally it could be executed in the early morning or late afternoon and kept shaded in the field to control the respiration rate that can affect the quality of produce. Produce must be handled gently to avoid from mechanical damage such as bruises, decay, skin damage and other injuries that would lead to moisture loss. Such damage could be prevented by applying appropriate postharvest handling such as harvest at correct maturity, handle produce in dry condition and use of bins equipped with internal padding.

The quality of produce cannot be improved after harvest, only maintained for the expected shelf life of the commodity. Harvesting time could vary dependent on market demand. Time of harvesting depending on various reasons. Some types of produce are more delicate when harvested earlier, while some of them have the option to harvest later to achieve optimum quality and maturity. Three factors during harvesting activities that influence the quality of produce are right harvesting time, at peak maturity and less damage towards produce.

Packinghouse Operation

A packinghouse is often a physical structure equipped with a few equipment such as conveyor, packaging facilities, washing systems and pre-cooling facilities. Operations and practices implemented in the packinghouse are dependent on demand and market requirements. Produce for domestic market is subjected to simple practices as produce may undergo a variety of handling practices, from dumping, sorting, trimming, washing, grading and packaging. It is important to locate packinghouse at the right area. Access to the field and market area, adequate space for vehicle to enter and leave, and accessible for workers are three aspects to take into consideration. In the simplest construction of packinghouse, produce from nearby farm is delivered in containers. The packers at the packinghouse then execute a variety of practices from sorting, grading, sizing and packing the produce before placing it directly into proper transportation containers. Person in charge in these practices has to be knowledgeable regarding maintenance of produce safety and quality, grade and size requirements and packing procedures. Sizing and packaging is still commonly practice in packinghouse. Person in charge for sizing practice must be knowledgeable in selecting desired size because of higher price the producer could get for particular size of produce. Packaging plays an important role in the maintenance of quality attributes of fresh produce. Various packaging materials used to wrap fresh produce to provide cushioning which then can avoid them from any possibilities of adverse effects (biological and mechanical damage) during handling, distribution and could also from environment. The mechanical barrier properties imply how well the packaging material can prevent the product from mechanical injuries during its distribution and storage life. In developing country, bamboo basket is commonly used for transportation but to let produce has direct contact with bamboo is not the best practice. Returnable plastic container then has been introduced to all players along postharvest supply chain to replace bamboo baskets for field handling practices.

Small-scale producers have the option to harvest earlier, when vegetables are more delicate and valuable; harvest later, when fruits are at a riper, more flavorful stage; or harvest more often (taking advantage of multiple harvests to gather produce at its optimum stage of maturity). All these options can lead to higher profits due to the higher value of the produce you have to offer for sale.

One of the most common mistakes growers make is to harvest fruit crops too early, when they are under-ripe and have not yet developed their full flavor. Some vegetables, if allowed to grow large, will be too fibrous or full of seeds for good eating quality. With many horticultural crops, if you harvest all at once you are sure to have many items that are either under-mature or over-mature. Using a maturity index as a standard will greatly reduce pre-sorting losses. For some crops this involves using a refractometer to measure sugars or a penetrometer to measure firmness.

Mechanical damage during harvest can become a serious problem, as injuries predispose produce to decay, increased water loss and increased respiratory and ethylene production rates leading to quick deterioration. In general, harvesting by machine will cause more damage than harvesting by hand, although some root crops can be severely damaged by careless hand digging. The containers used by pickers in the field should be clean, have smooth inside surfaces and be free of rough edges. Stackable plastic crates, while initially expensive, are durable, reusable and easily cleaned. If baskets must be used, they should be woven "inside out" with the stubs of the beginning and end of each cane on the outside of the basket.

Manual harvesters should be well trained in the proper way to harvest the crop to minimize damage and waste, and should be able to recognize the proper maturity stage for the produce they are handling. Pickers should harvest with care, by snapping, cutting or pulling the fruit or vegetable from the plant in the least damaging manner. The tips of knives should be rounded to minimize inadvertent gouges and excess damage to perennial plants. Knives and clippers should always be well sharpened. Pickers should be trained to empty their picking bags and baskets with care, never dumping or throwing produce into field containers. If harvesters pick directly into large bulk bins, produce can be protected from bruising by the use of a de-accelerating chute fashioned from canvas. Vented, stackable field containers should be kept clean and smooth.

Exposure to the sun should be avoided as much as possible during and after harvest, as produce left out in the sun will gain heat and may become sun-burned. Produce exposed to sunlight can soon become 4 to 6° C (7 to 11° F) warmer than air temperature. Field bins should be placed in the shade or loosely covered (for example with light colored canvas, leafy plant materials, straw or an inverted empty container) if delays are expected in removing them from the field. Night or early morning harvest is sometimes an option for harvesting produce when internal temperatures are relatively low, reducing the energy needed for subsequent cooling. Latex flow is often lower later in the morning than it is at dawn for crops such as mango and papaya, so harvesting as late in the morning as possible can reduce later efforts required to clean the produce before packing. Also, citrus fruits should not be harvested early in the morning when turgid because of their greater susceptibility to the release of essential oils from the flavedo oil glands that cause oil spotting (green spots on yellow and orange citrus fruits after degreening).

Directly following harvest, when produce is prepared for marketing, cooling is essential. Cooling (also known as "pre-cooling") is the removal of field heat directly after harvest, before any further handling. Any delays in cooling will shorten postharvest life and reduce quality. Even produce undergoing repeated cooling and warming deteriorates at a slower rate than produce that has not been cooled.

Rough handling during preparation for market will increase bruising and mechanical damage and limit the benefits of cooling. Roads between the field and the packinghouse should be graded and free from large ruts, bumps and holes. Field boxes must be well-secured during transport and, if stacked, not overfilled. Transport speeds must be suited to the quality and conditions of the roads, and truck and trailer suspensions kept in good repair. Reduced tire air pressure on transport vehicles will reduce the amount of motion transmitted to the produce.

Any practice that reduces the number of times the produce is handled will help reduce losses. Field packing (selection, sorting, trimming and packaging of produce at the time of harvest) can greatly reduce the number of handling steps the produce must undergo before marketing. Small carts or small mobile field packing stations can be designed to be moved along with the packers and to provide shade for packing operations.

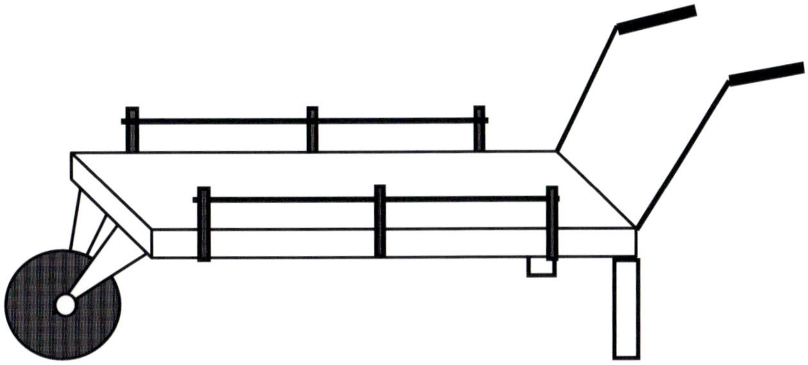

HAND CART: Used for field packing. Harvesters place an empty carton on the cart, then pick, sort, grade and pack directly into the container. Filled cartons are moved immediately into coolers or keep in a well-shaded area.

Maturity Standards

Maturity standards have been determined for many fruit, vegetable and floral crops. Harvesting crops at the proper maturity allows handlers to begin their work with the best possible quality produce. Produce harvested too early may lack flavor and may not ripen properly, while produce harvested too late may be fibrous or overripe. Pickers can be trained in methods of identifying produce that is ready for harvest. The following table, provides some examples of maturity indices:

Index	Examples
Elapsed days from full bloom to harvest	Apples, pears
Mean heat units during development	Peas, apples, sweet corn
Development of abscission layer	Some melons, apples, feijoas
Surface morphology and structure	Cuticle formation on grapes, tomatoes Netting of some melons Gloss of some fruits (development of wax)
Size	All fruits and many vegetables
Specific gravity	Cherries, watermelons, potatoes
Shape	Angularity of banana fingers Full cheeks of mangos Compactness of broccoli and cauliflower
Solidity	Lettuce, cabbage, brussels sprouts
Textural properties	
Firmness	Apples, pears, stone fruits
Tenderness	Peas
Color, external	All fruits and most vegetables
Internal color and structure	Formation of jelly-like material in tomato fruits Flesh color of some fruits
Compositional factors	
Starch content	Apples, pears
Sugar content	Apples, pears, stone fruits, grapes
Acid content, sugar/acid ratio	Pomegranates, citrus, papaya, melons, kiwifruit
Juice content	Citrus fruits
Oil content	Avocados
Astringency (tannin content)	Persimmons, dates
Internal ethylene concentration	Apples, pears

Vegetables are harvested over a wide range of maturities, depending upon the part of the plant used as food. The following table provides some examples of maturity indices of vegetable crops.

Crop	Index
Root, bulb and tuber crops	
Radish and carrot	Large enough and crispy (over-mature if pithy)
Potato, onion, and garlic	Tops beginning to dry out and topple down
Yam bean and ginger	Large enough (over-mature if tough and fibrous)
Green onion	Leaves at their broadest and longest
Fruit vegetables	
Cowpea, yard-long bean, snap bean, batao, sweet pea, and winged bean	Well-filled pods that snap readily
Lima bean and pigeon pea	Well-filled pods that are beginning to lose their greenness

Okra	Desirable size reached and the tips of which can be snapped readily
Upo, snake gourd, and dishrag gourd	Desirable size reached and thumbnail can still penetrate flesh readily (over-mature if thumbnail cannot penetrate flesh readily)
Eggplant, bitter gourd, chayote or slicing cucumber	Desirable size reached but still tender (over-mature if color dulls or changes and seeds are tough)
Sweet corn	Exudes milky sap from kernel if cut
Tomato	Seeds slipping when fruit is cut, or green color turning pink
Sweet pepper	Deep green color turning dull or red
Muskmelon	Easily separated from vine with a slight twist leaving clean cavity
Honeydew melon	Change in fruit color from a slight greenish white to cream; aroma noticeable
Watermelon	Color of lower part turning creamy yellow, dull hollow sound when thumped
Flower vegetables	
Cauliflower	Curd compact (over-mature if flower cluster elongates and become loose)
Broccoli	Bud cluster compact (over-mature if loose)
Leafy vegetables	
Lettuce	Big enough before flowering
Cabbage	Head compact (over-mature if head cracks)
Celery	Big enough before it becomes pithy

Processing of Crops

When conditions are not suitable for storage or immediate marketing of fresh produce, many horticultural crops can be processed using simple technologies. There are many processing methods that can be used by small-scale handlers, including drying, fermenting, canning, freezing, preserving and juicing. Fruits, vegetables and flowers can all be dried and stored for use or sale in the future. Fermentation is popular throughout the world as a food preservation method, and over 3,500 individual fermented foods have been described by Campbell-Platt. Fruits and vegetables can be canned or frozen, and fruits are often preserved in sugar or juiced.

Processed products must be packaged and stored properly in order to achieve their potential shelf life of up to one year. Dried products must be packaged in air-tight containers (glass or plastic bottles or sealed plastic bags). Canned and bottled products must be adequately heat processed using high quality containers that provide good seals. Dried and canned or bottled products are best stored in a cool, dark place.

Postharvest handling, transport and marketing of processed products can be much simpler and less costly than for fresh products, since refrigeration is unnecessary. Dried products take up much less space than their fresh equivalents, further reducing transport and storage costs.

25 pounds of:	Yields this much dried product:
Fruits	4 lb
Carrots or beets	3 lb
Celery, cabbage or tomatoes	1.5 lb
Onions or zucchini	2.5 lb

Processing Equipment

A catalog of postharvest processing equipment is available from Intermediate Technology Publications. Included are driers, storage containers, cleaners, hand mills, power mills, shellers, decorticators (seed removers), oil processing equipment, fruit presses, and root crop cutters/graters. Some examples are shown below:

Two-man cassava grater and Four-bladed root chopper:

Hand-operated fruit press and Cherry pitter:

Preparation for Processing

Some produce requires blanching before freezing or drying. Blanching by boiling water bath or in steam ends certain enzymatic reactions in the product and helps retain color and flavor after processing. Always rinse blanched produce under very cold water or dip blanched produce into an ice water bath to stop the cooking process and quickly bring the temperature down.

Blanching times for selected commodities {use one gallon of water per pound (8 liters per kg) of produce}. Add one minute for each 2000 ft altitude if you live at elevations over 4000 ft.

Commodity	Blanching time in boiling water (minutes)
Broccoli	3
Green Beans	3
Cabbage (wedges)	5
Carrots	5
Cauliflower	3 (add 4 teaspoons of salt)
Collard greens	3
Corn (sweet)	7
Eggplant -	4 (add 1/2 cup lemon juice)
Leafy greens	2
Mushrooms	3 to 5
Peas	2
Potatoes (new)	4 to 10
Pumpkin	2 to 3 or until soft
Sweet potatoes	15 to 20 or until soft
Zucchini/summer squash	3

Fruits such as apples, pears, peaches and apricots are sometimes treated with sulphur being dried. Sulfuring {burn one tablespoon of sulfur powder per pound (12g per kg) of fruit} or sulphiting (dip fruit in a 1% potassium metabisulfite solution for one minute) helps prevent darkening, loss of flavor and loss of vitamin C.

Sulfur has been a source of allergic reactions in some people, so packages of sulphured product should always be clearly labelled. Vitamin C can be used as an alternative pre-treatment to prevent browning during the drying process. Use 30 ml ascorbic acid powder in one litre (or 2 tablespoons in one quart) of lukewarm water. Slice the fruits directly into the solution, remove with slotted spoon, drain well and pat dry.

For best results when drying fresh produce, fruits should be sliced or quartered, and vegetables should be thinly sliced, chopped or diced. Solar drying of fruits will take 2 to 3 days or longer, while most chopped or diced vegetables will dry in 1 to 2 days.

Sulfuring times for selected fruits:

Commodity	Sulfuring Time for Quartered Fruits	Sulfuring Time for Halved Fruits
Apples	45 minutes	
Apricots	2 hours	3 hours
Cherries	20 minutes	30 minutes
Nectarines	2 hours	3 hours
Peaches	2 hours	2 to 3 hours
Pears	2 hours	4 to 5 hours

A low cost sulfuring box can be constructed from a large cardboard box that is slashed in several places to allow adequate ventilation. Trays for drying be stacked using bricks and wooden spools as spacers. The trays must be made completely of wood, since sulfur fumes will corrode metal The entire assembly must be located out of doors, preferably on bare soil. Use one tablespoon of sulfur powder per pound (35 mls per kg) of fruit. Place the sulfur in a container well away from the side of the box since it will become quite hot. Seal the bottom edges of the box with soil.

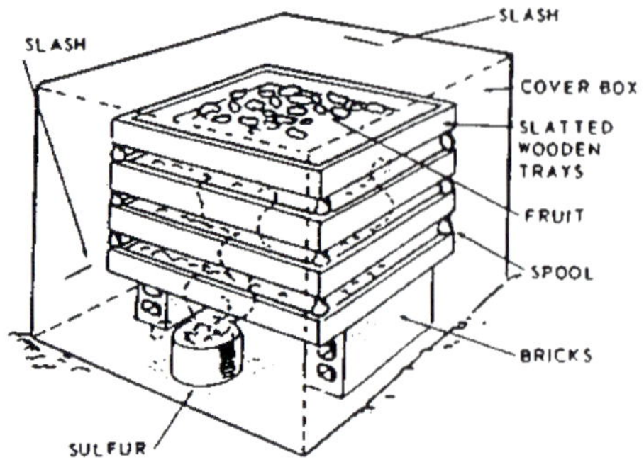

Wooden box:

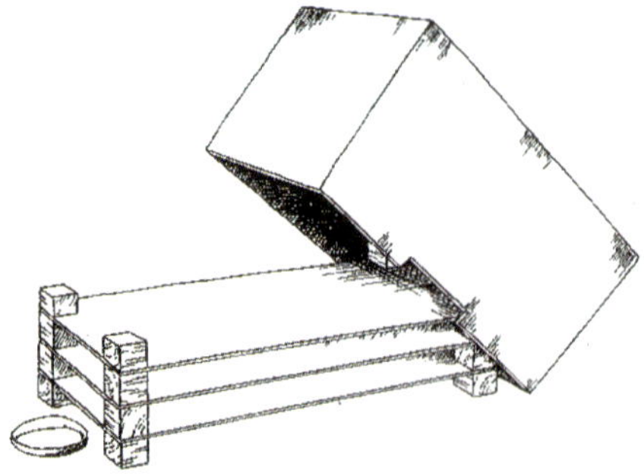

Solar Drying

Horticultural produce can be dried using direct or indirect solar radiation. The simplest method for solar drying is to lay produce directly upon a flat black surface and allow the sun and wind to dry the crop. Nuts can be dried effectively in this way.

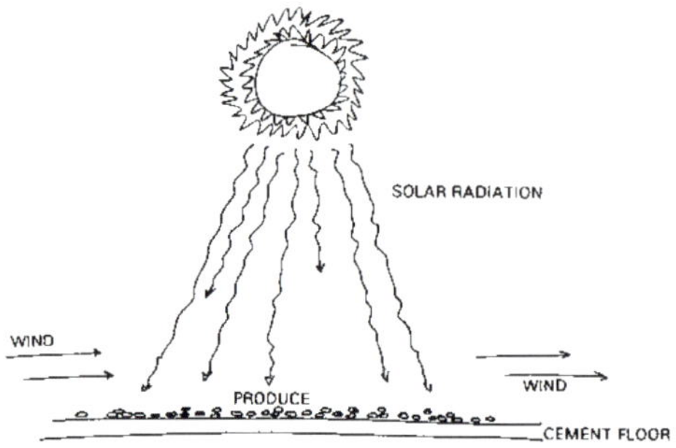

Simple direct driers can be made from trays of screening material propped upon wooden or concrete blocks to allow air to circulate under the produce. A layer of cheesecloth can be draped loosely over the produce, protecting it from insects and birds while drying. Check the produce each day, and move it under cover if rain threatens.

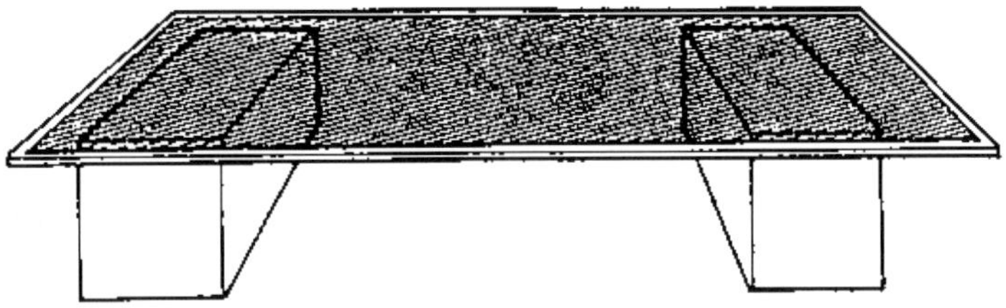

A simple method for solar drying is to construct a raised platform from wood and cover the frame with loosely woven mats. In the illustration below, sliced fresh tomatoes are being dried in direct sunlight on straw mats. Air can pass over and below the produce, speeding drying and reducing losses due to overheating.

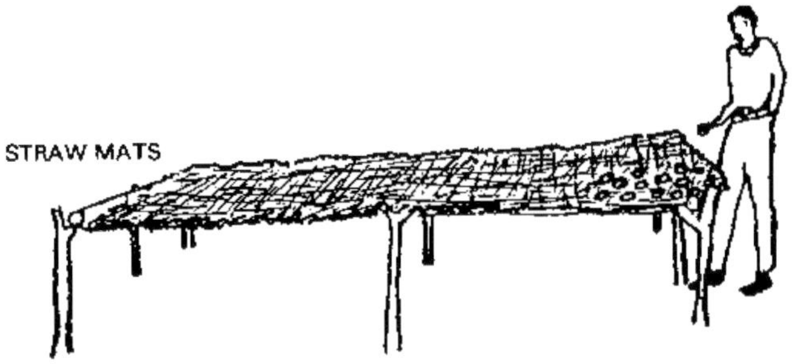

Aluminum foil can be utilized to reflect the sun onto the drying trays. The example shown below uses a plastic sheet to trap some of the heat and speed drying time.

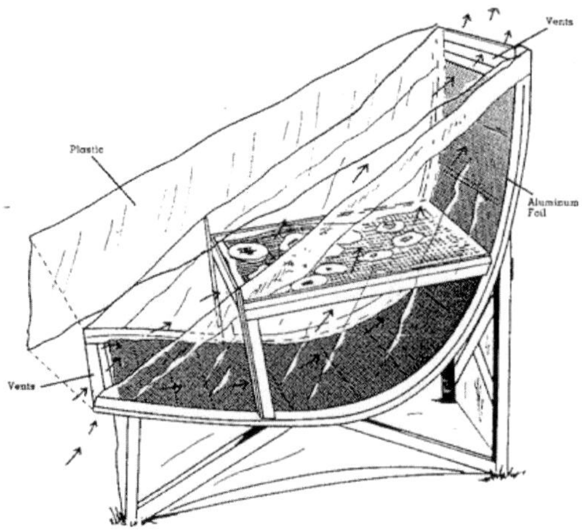

All the trays, screens or mats used for drying produce must be kept clean. Trays made from stainless steel, plastic or nylon are much easier to keep clean than are wooden trays. Some fruits juices will adhere to the surfaces with each handling. This leads to an accumulation of dirt and buildup of mold that can contaminate and affect the appearance of the dried product. Use a strong detergent and a still brush to scrub the trays, screens or mats clean. Allow them to dry in the sun before using them to dry product.

A good material for settling dust on roads, paths or earthen floors is calcium chloride. When spread on the earth it absorbs moisture from the air and keeps the soil moist. Rake the calcium chloride into the surface at a rate of ½ pound per square yard.

To reduce mold growth on trays, screens and mats during the off season, wash and dry them thoroughly, then store them in a well ventilated area.

In order to improve the efficiency of drying, some sort of structure must be used to capture solar radiation. Various types of solar driers have been developed and are illustrated below:

Drier type	Description	Schematic view of typical example
Direct cabinet	Drying Chamber is glazed and no separate solar collector is used.	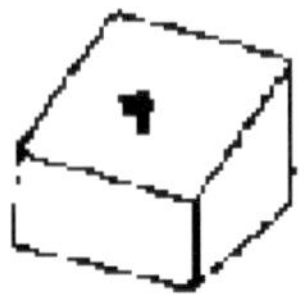
Indirect cabinet	Solar collector is used which is separate from the drying chamber, which has no transparent surfaces.	
Mixed mode or Hybrid cabinet	Drying chamber is partially or fully glazed, and a separate solar collector is used.	
Tunnel	Usually a hoop framed structure with one or two layers of film plastic glazing. Usually a direct drier, but can be indirect if black plastic film is used for the inner layer of glazing.	
Low tunnel	Direct drier similar to the above but built close to the ground, and usually able to hold only one layer of produce.	
Tent	Direct drier with straight rather than curved frame members.	
Bin	Any drier, but usually indirect with forced convection air flow, which can dry deep layers (typically 300 mm or more) of the produce.	

More complex models of solar driers have glass or clear plastic windows that cover the produce, providing some protection from insects while capturing more of the heat of the sun.

Direct solar drier:

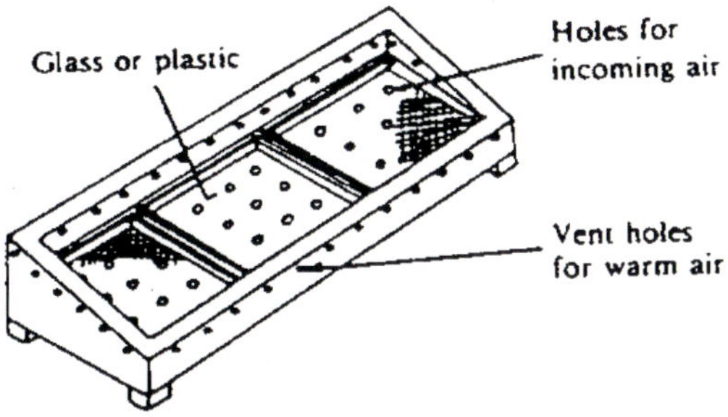

Indirect driers are constructed so the sun shines upon a solar collector (a shallow box, the insides painted black, topped with a pane of glass) heating air which then moves upward through a stack of four to six trays loaded with produce.

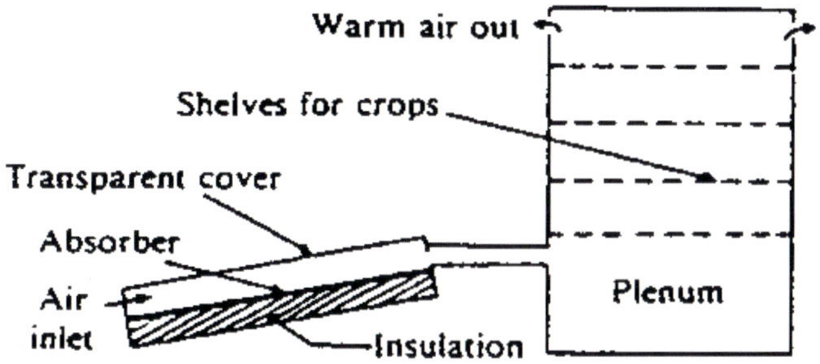

Forced-air Dehydrators

Nut crops can be dried in bulk using a dehydrator that combines a steady stream of air with an external source of heat. The plenum chamber below the produce is covered with a floor of perforated sheet metal or wooden slats. A fan located between the furnace and the plenum chamber moves the hot air through the drying produce.

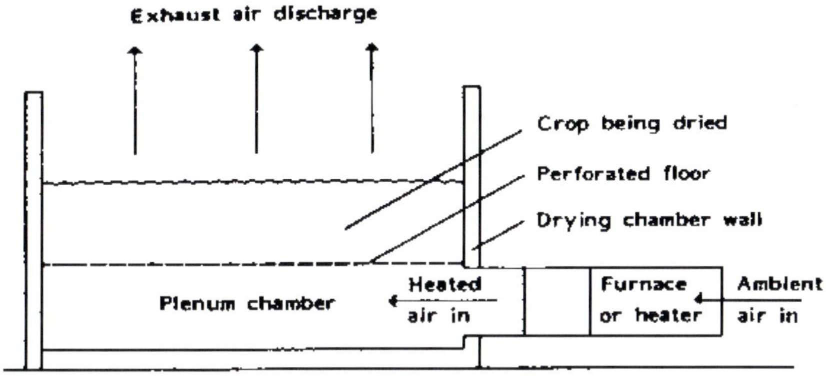

Oil-burning Dehydrators

The batch-dryer illustrated below is constructed of wood, has an axial type fan and burns kerosene or diesel oil. A wide variety of dryers are available from manufacturers around the world.

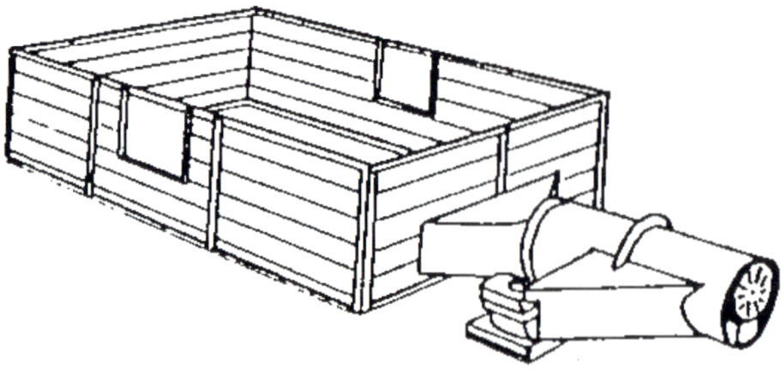

Two types of dehydrators are commonly used for drying small volumes of nut crops. A wagon with a perforated floor can be transported from the field and connected to a portable burner batch drying. A stationary "pot-hole" dehydrator is designed to move heated air along a plenum under a fixed platform Individual bins of nuts are placed upon the platform and are dried as heat rises up through the perforated floor.

Wagon Dehydrator:

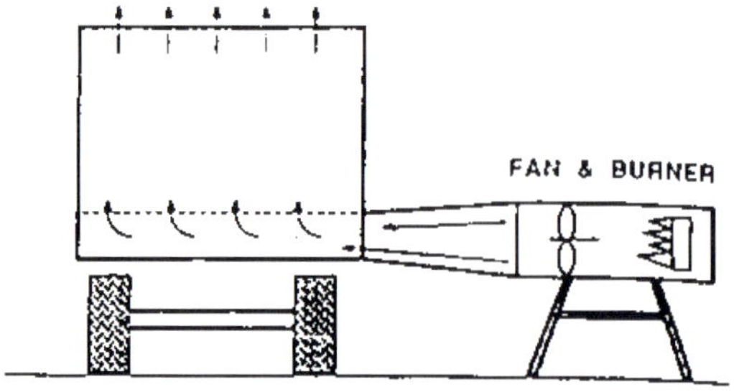

Pot-Hole Dehydrator:

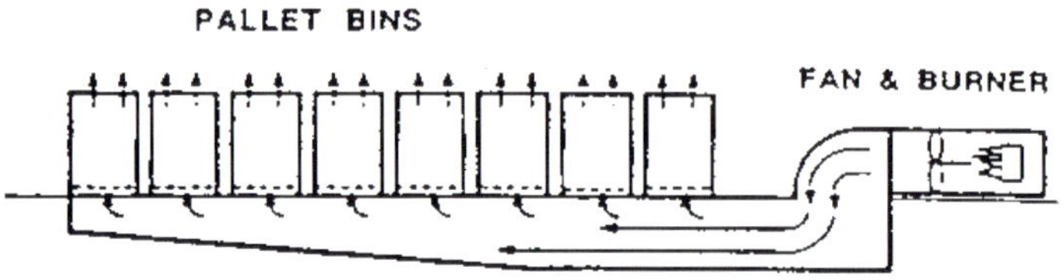

Electric Dehydrators

A simple electric dehydrator can be constructed using plywood, sheet metal a small fan, five household light bulbs with porcelain mounting fixtures and some screening material. The drier shown below is 32 inches long by 21 inches wide and 30 inches high, and contains racks for five trays. The fan and the sheet metal lining the bottom compartment help conduct heat upward through the box.

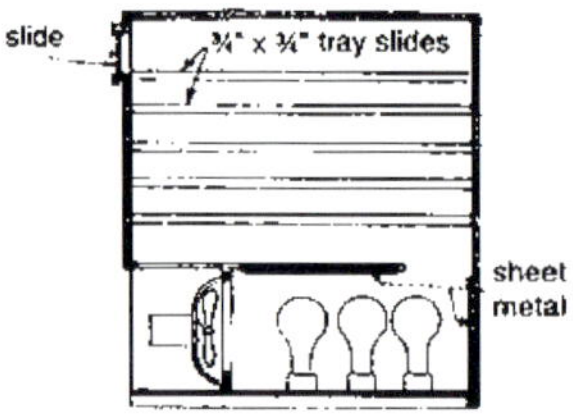

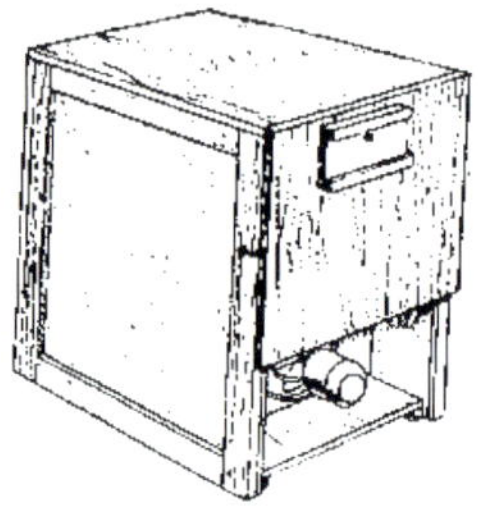

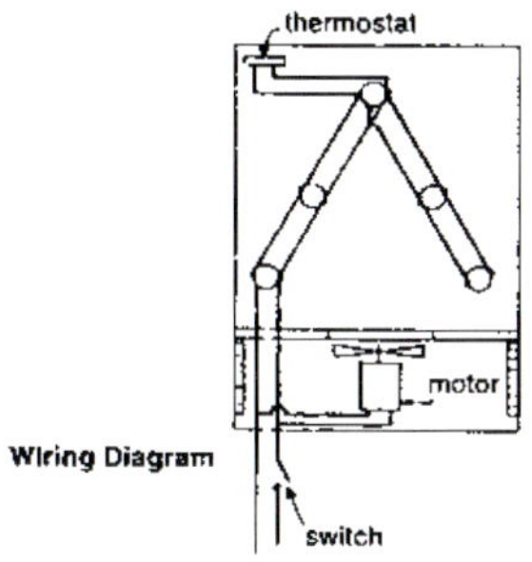

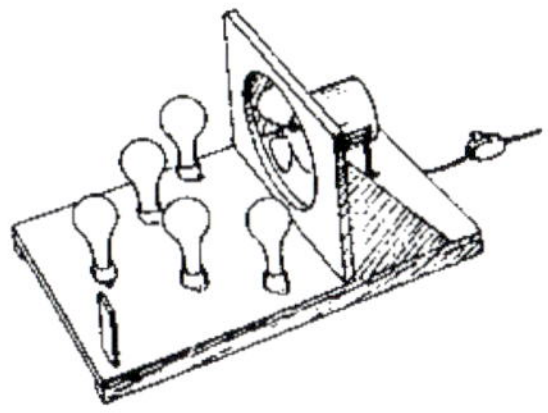

Oven Drying

Fruits and vegetables can be dried in a home oven if the oven can be run at a low temperature. Place the prepared produce on baking or metal screen trays, set the oven temperature at 140 °F and leave the door ajar (2 to 4 inches). Drying time can be reduced if ventilation is increased by using a small fan placed outside the oven.

Drying Flowers

Cut flowers can be air dried by hanging upside down or while supported by chicken wire. Certain flowers will look more natural if left standing in a vase while they dry. Anthurium dries best if left to dry very slowly. Cut the stems at a sharp angle, and place the flowers into a vase containing two inches of water. In all cases, flowers should be left to air dry in a dark, well area.

Flowers that dry best if left standing: strawflower, delphinium, larkspur, okra pods

Flowers that dry best while hanging upside down: chysanthemum, amaranthus, African daisy, statice, marigold.

African daisy supported on a screen of chicken wire:

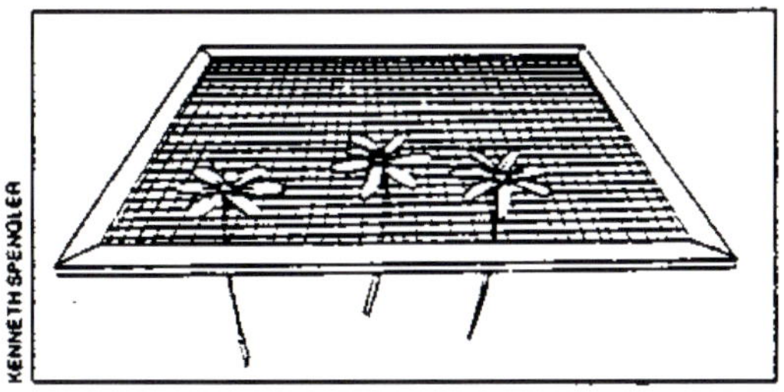

Cut flowers can be dried quickly and easily in sand or silica gel. Sand used for drying flowers should be clean, smooth and the finer the better. Starting with one inch of sand in a container, place the flower to be dried on the sand and gently cover the entire flower with more sand. The container should be left uncovered and flowers should be dried in about three weeks. Flowers that dry well in sand are shasta daisy, lily-of-the-valley, cosmos, dahlia, sweet william carnation, stock, freesia and narcissus.

Drying flowers in sand:

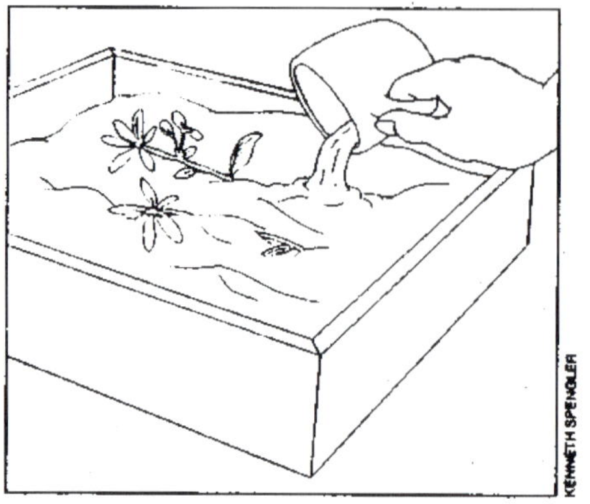

Silica gel is relatively expensive but reusable if heated to dry out the gel between uses. To use, cover the flower as with sand, then tightly seal the container. Check for drying in two to three days. Silica gel is especially useful for drying fragile plants and flowers with delicate colors.

Flowers that dry best in silica gel are allium, anemone, cornflower, roses, tulip and zinnia.

Canning

Two types of canners are commonly used to process horticultural crops. The first is a water bath canner, which is a large pot with a loose cover and a rack to hold jars off the bottom The pot should be deep enough to cover the canning jars by one to two inches and still have another inch of space to allow brisk boiling. The diameter of the pot should be no more than four inches wider than the diameter of the stove's burner to ensure even heating.

Acidic foods such as fruits, tomatoes, pickles and relishes, and high sugar foods such as jams, jellies, syrups and marmalades can be safely processed using a boiling water bath.

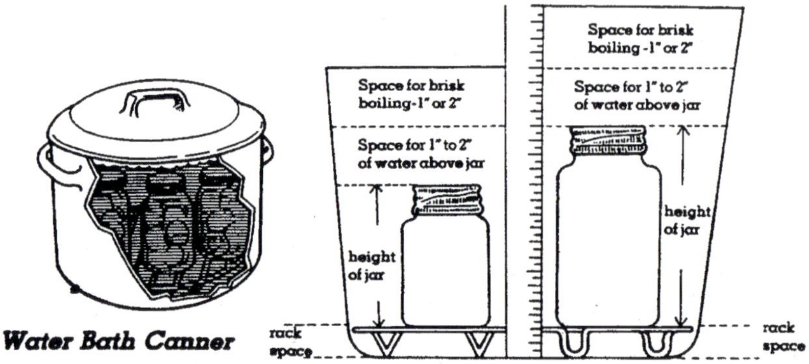

A pressure canner is recommended for processing low acid foods such as vegetables. A pressure canner is a specially made heavy pot with a locking lid, an inner rack and a steam vent in the kid. The vent can be adjusted using a weight, value or screw, depending on the type of canner. A pressure gauge registers the air pressure inside the canner. A dial gauge gives a reading of the actual pressure, while weighted gauges will rock gently when the canner is at the proper pressure. Ten pounds of pressure at 115° C (240° F) is recommended for canning vegetables.

Pressure Canners and Dial Gauge:

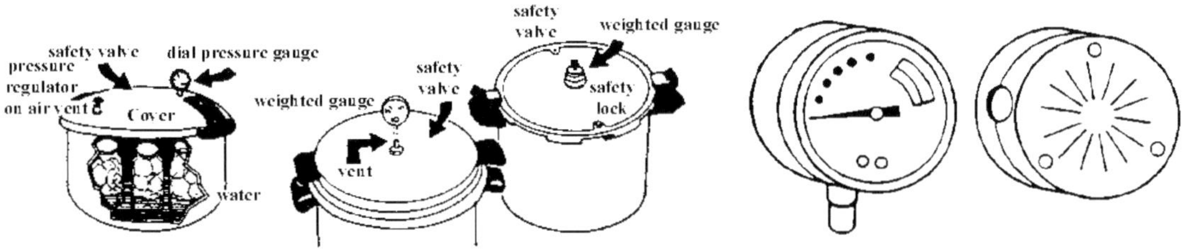

Weighted Gauges:

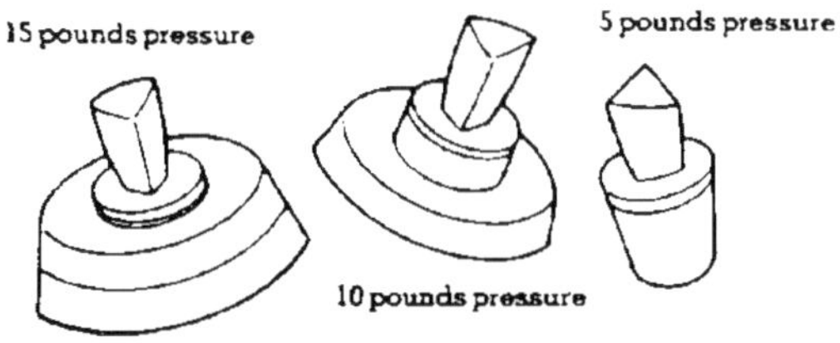

There are three types of glass canning jars used for processing horticultural crops. The ball type jar and the zinc capped jar both require rubber rings as seals. These can sometimes be difficult to obtain, but if locally available, make excellent containers. Currently the canning jar with a two-pieced lid is the most commonly used container.

No matter which jar is used, when filling containers, it is important to leave a small amount of headspace to allow for expansion of the food while processing. If a jar is filled too full, it may explode. If too much headspace is left, the food may spoil, since all the extra air may not be driven out during processing.

Canning jars and lids, L to R: Ball type jar, zinc lid and two-piece:

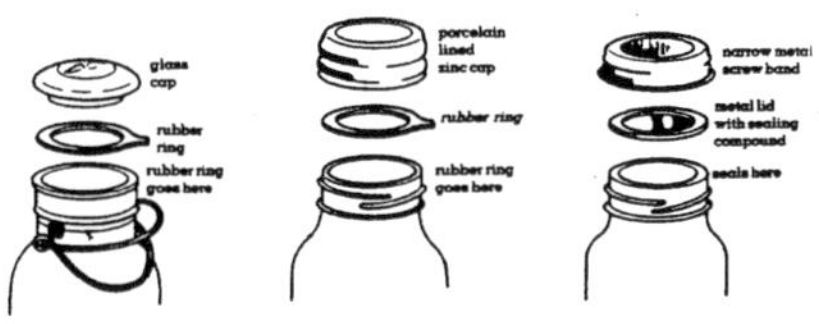

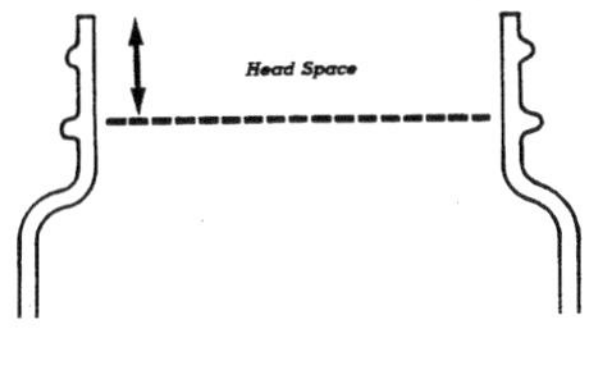

Juicing

Fruits

To process tomatoes or fruits to juices, fruits are simmered in water or their own juice in a stainless steel, glass or enamelware pot. When tender, the product is cut into pieces and pressed through a food mill, colander or several layers of cheesecloth. Sugar or lemon juice can be added, to taste.

The juices must then be either frozen or canned for storage. Juices can be frozen in jars or freezer containers (leave 1/2 inch headspace). Most fruit juices can be canned in a boiling water bath for 20 minutes, but apple and grape juices can be processed in hot water (82° C or 180° F) for 30 minutes.

Vegetables

Vegetables should be chopped or shredded, then simmered for 45 to 50 minutes until mushy. The juice can then be pressed or strained from the vegetable pulp, and frozen or canned. Canning vegetable juices requires processing at ten pounds of pressure in a pressure canner. Pints should be processed for 55 minutes, and quart jars for 85 minutes.

Other Methods of Processing

Freezing

Most vegetables should be blanched before freezing to prevent loss of flavor and color during storage. Freezing temperatures are best set between -21 to -18° C (0 to 5° F).

Packages for freezing should be moisture proof and vapor proof and contain as little air as possible to prevent oxidation during storage. Heavy plastic bags, heavy aluminum foil, glass freezer jars and waxed freezer cartons all make good containers.

Jellies, Jams and Preserves

Making jams, jellies and other high sugar preserves requires a balance of fruit, acid, pectin and sugar for best results. Under-ripe fruits contain more pectin than ripe fruits, and apple juice is a good source of natural pectin. If fruits are low in acid, lemon juice can be added to the mixture of fruit and sugar. Cane or beet sugar is better for making preserves than honey or corn syrup.

To preserve fruits, cook on medium heat until the mixture "sheets" from a spoon. Avoid overcooking since this will lower the jelling capacity of the mixture. Pour into containers and seal with paraffin wax (jellies only). The other preserves should be processed in a boiling water bath for five minutes.

Fermentation

When lactic acid bacteria in foods convert carbohydrates to lactic acid, food is preserved by the resulting low pH. Sauerkraut (cabbage) and wine (grapes) are two examples of thousands of fermented foods made around the world.

Acidification

Pickling is a simple processing method that can be used with many types of fruits and vegetables. Brine solution (9 parts cider or white vinegar, 1 part non-iodized salt, 9 parts water, plus flavorings and spices) is poured over the product into glass canning jars (leave 1/2 inch headspace). Brined pickles are sealed and left at ambient temperature for three or more weeks, while fresh pack pickles are processed in a boiling water bath for 10 minutes.

Harvesting Practices

Harvesting practices should cause as little mechanical damage to produce as possible. Gentle digging, picking and handling will help reduce crop losses.

Pick carefully to avoid damage:

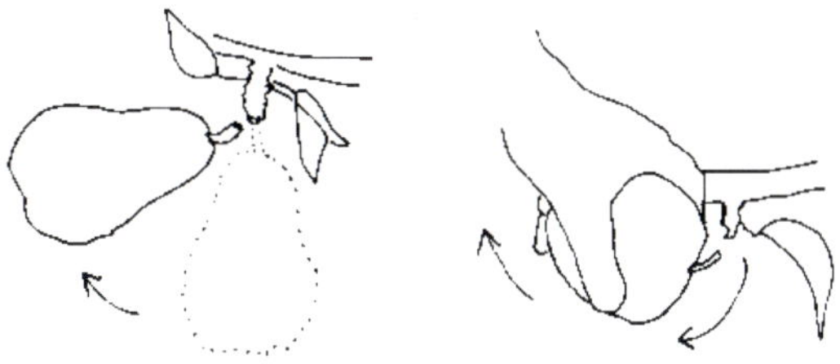

For some crops, a natural break point forms at the junction of the stem and the stalk when produce is mature. Harvesters should grasp the product firmly but gently and pull upward as illustrated below. Wearing cotton gloves, trimming fingernails, and removing jewelry such as rings and bracelets can help reduce mechanical damage during harvest.

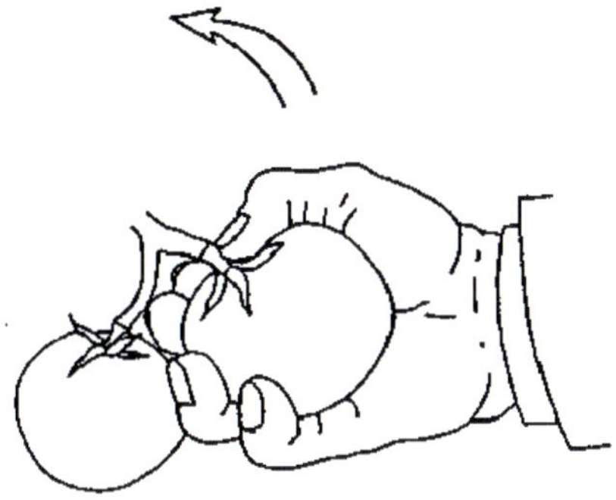

If a small amount of leafy vegetables are being harvested for home use or for sale at a nearby roadside or farmers' market, a small tub of cold water can be useful for cooling the produce. The tub can be brought directly to the field and used by the picker as a field container. Clean water should be used with each lot of produce. Chilling leafy vegetables by using cold water at harvest will help maintain quality and prevent wilting.

Harvesting Containers

Picking baskets, bags and buckets come in many sizes and shapes. Buckets are better at baskets in protecting produce, since they do not collapse and squeeze produce.

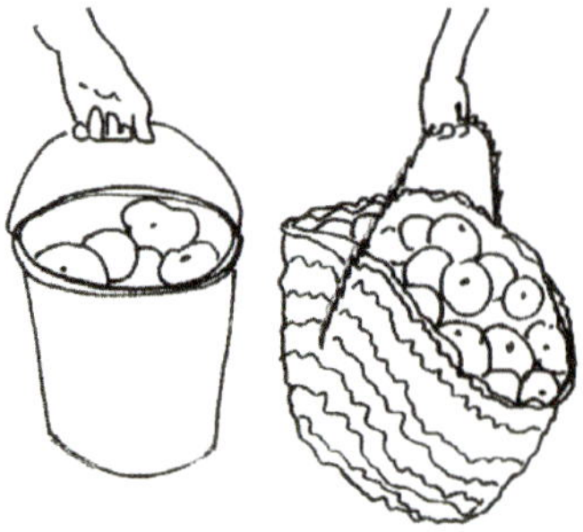

These harvesting containers can be made by sewing bags with openings on both ends, fitting fabric over the open bottom of ready-made baskets, fitting bags with adjustable harnesses, or by simply adding some carrying straps to a small basket.

Plastic crates are relatively expensive but are durable, reusable and easy to clean. When empty, they can be nested to save space in storage or transport. When filled they can be stacked if every other crate is turned in the direction opposite to the one below.

Stackable, reusable plastic crates:

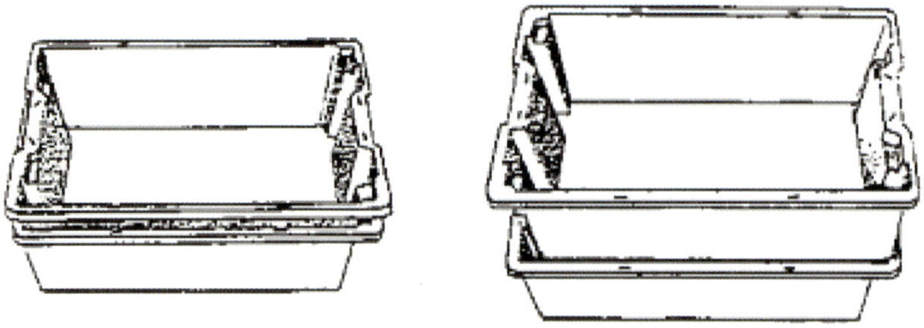

If plastic crates are well vented along the sides and bottom, they can also be used to wash and/or cool produce after harvest.

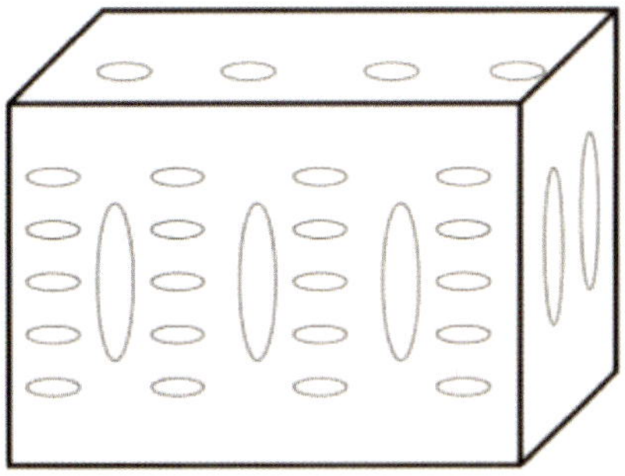

Harvesting Tools

Some fruits need to be clipped or cut from the parent plant. Clippers or knives should be kept well sharpened. Penduncles, woody stems or spurs should be trimmed as close as possible to prevent fruit from damaging neighboring fruits during transport.

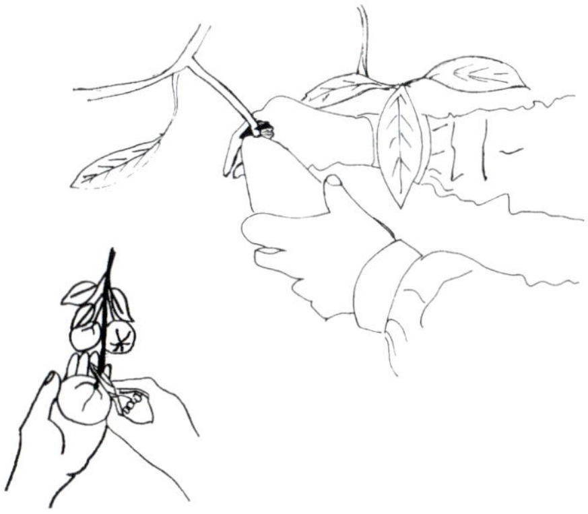

Pruning shears are often used for harvesting fruits, some vegetables, and cut flowers. A variety of styles are available as hand held or pole models, including shears that cut and hold onto the stem of the cut product. This feature allows the picker to harvest without a catching bag and without dropping fruits.

Straight bladed hand shears for fruits and flowers:

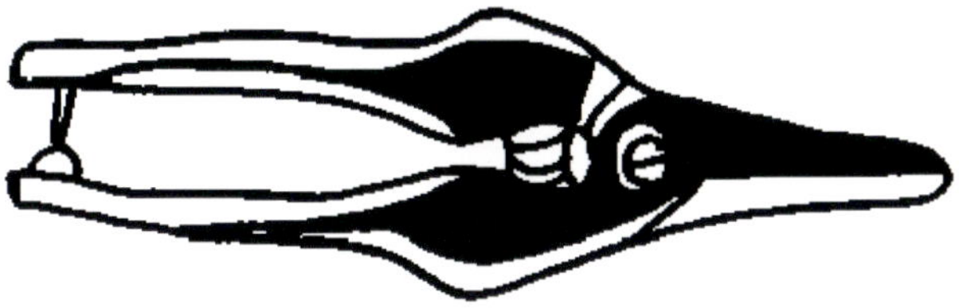

Thin curved blade for grapes and fruits:

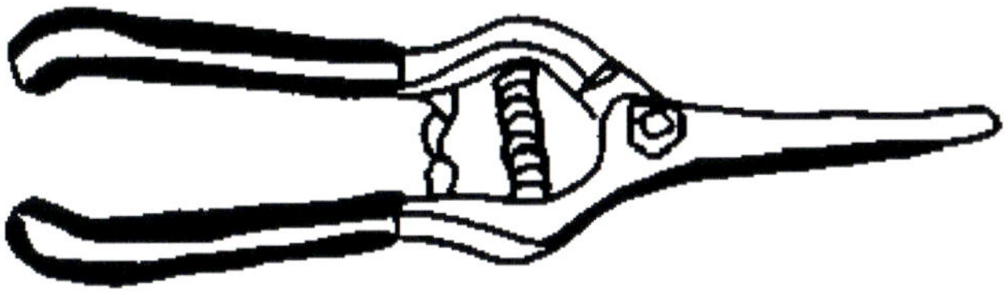

Cut and hold hand shears:

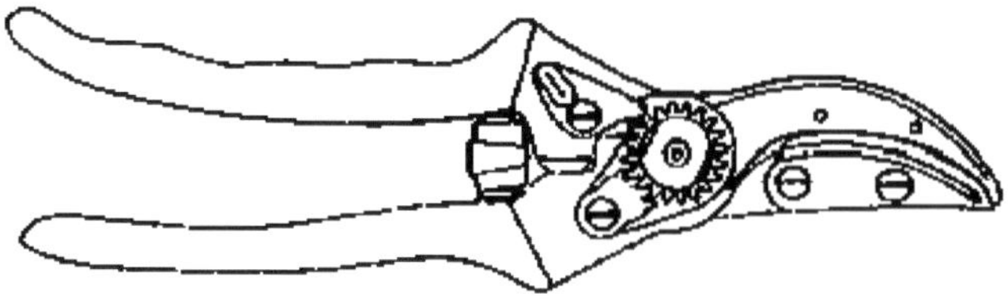

Clipper for citrus fruits:

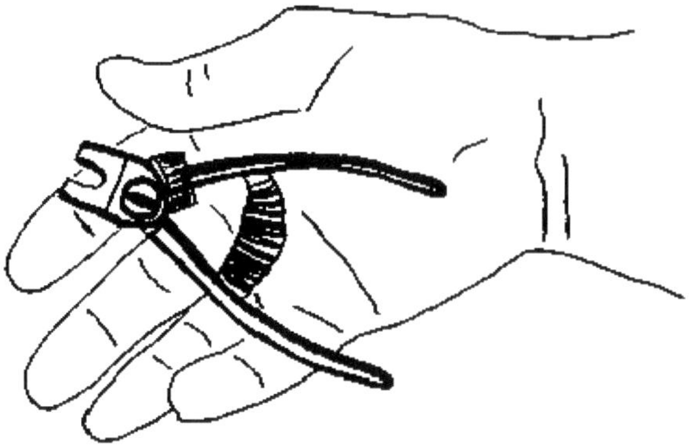

Pole mounted cut and hold picking shears:

Using a cutting tool attached to a long pole can aid picking of crops such as mangoes and avocados when the fruit is difficult to reach. Cutting edges should be kept sharpened and the catching bag should be relatively small The angle of the cutting edge and the shape of the catching bag can affect the quality of the fruit harvested, so it is important to check performance carefully before using any new tools.

Using a picking pole:

Picking poles and catching sacks can be made by hand or purchased from horticultural supply companies. The collection bags illustrated below were hand woven from strong cord or sewn from canvas. The hoop used as the basket rim and cutting edges can be fashioned from sheet metal, steel tubing or recycled scrap metal.

Hand Woven Collection Bag

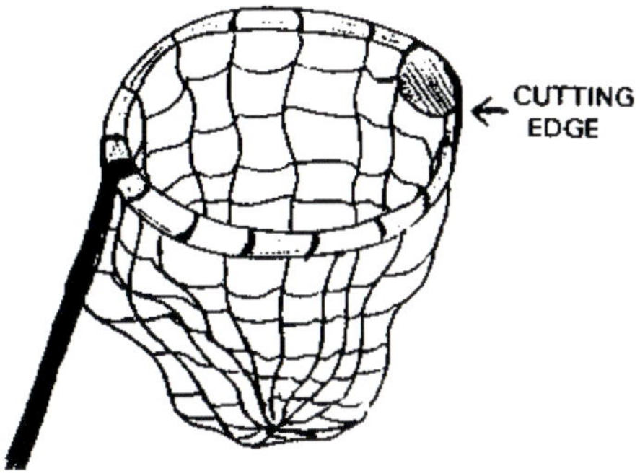

Canvas Collection Sack

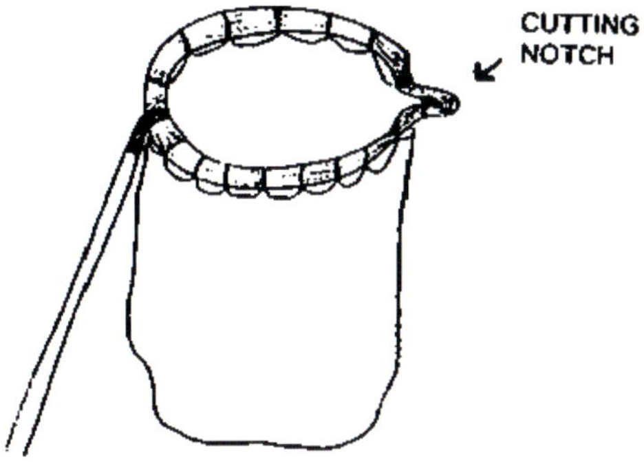

Fruit trees are sometimes quite tall and letting fruit fall to the ground when it is cut from the tree will cause severe bruising. If two pickers work together, one can clip or cut the fruit from the tree, and the other can use a sack to break its fall. The catcher supports the bag with his hands and one foot, catches the falling fruit, then lowers the far end of the bag to allow the fruit to roll safely to the ground.

Unlike most nut crops, pistachio nuts should not be knocked to the ground during harvest because of their open shells and relatively high moisture content. The harvesting practice illustrated below can be used with pistachios and olives with good results. Plastic sheeting or canvas tarpaulins are spread below the tree being harvested, and trees are mechanically shaken or hand knocked (the branches hit with mauls) until the nuts drop. In the illustration below, two harvesters are gathering a sheet covered with produce.

Field Packing

When crops are field packed the picker harvests and then immediately packs the produce after minimal handling Strawberries are generally field packed, since even a small amount of handling will damage these soft fruits. When lettuce is field packed, several wrapper leaves are left on the head to help cushion the produce during transport.

A small cart can help reduce the amount of bending and lifting the picker has to do during harvest. The carts shown below have a single wheel in front, and can be pushed along the row ahead of the picker.

Field packing strawberries:

Field packing lettuce:

A simple aid for field packers is a movable cart with a rack for boxes and a wide roof to provide shade. This small cart is designed to be pushed by hand along the outer edge of the field or orchard where harvest is taking place. It has been used to field pack table grapes, small fruits and specialty vegetables.

This larger cart for field packing is designed to be pulled by a small tractor into the field when the crop is harvested. This type of cart can be used for field packing many types of crops. The roof folds down for easy transport, and opens up to provide a wide area of shade for the packers and the commodity. The cart design can be modified as needed to suit various products and different operations.

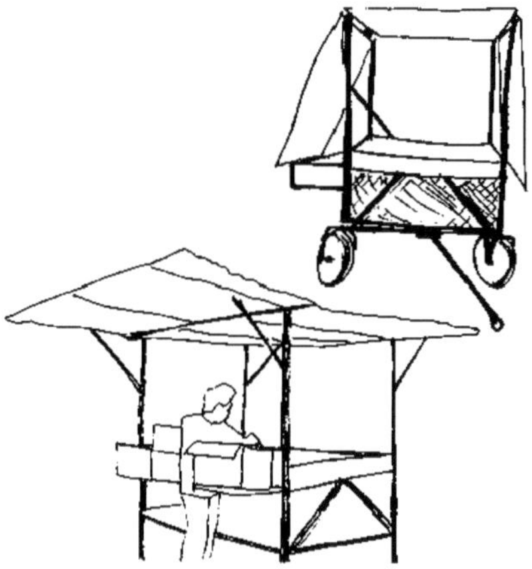

A self-propelled field pack system allows field workers to cut, trim, tie/wrap and pack in the field, thus eliminating the expense of operating a packing shed. In the illustration below, a fiat bed truck is moving along-side the field pack system, and packed produce is being loaded for transport.

Self propelled field pack system

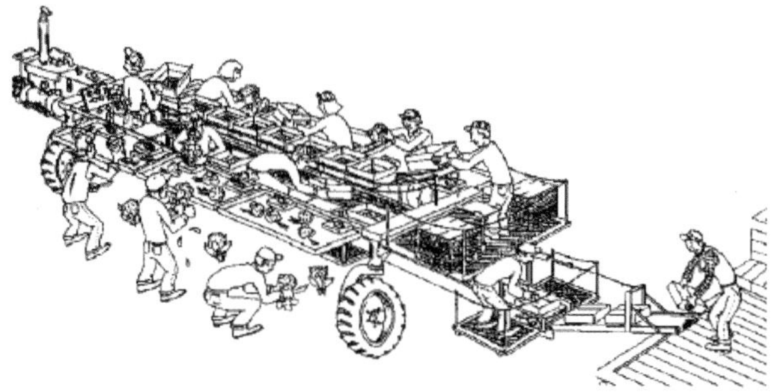

Options:

- Number of Worker positions per side
- Row Spacing
- Ground Clearance
- Can Harvest front, rear, or both

Transport to the Packinghouse

When crops are harvested at some distance from the packinghouse, the produce must be transported before packing. The gravity driven conveyor system for bananas illustrated below provides an example of how handling can be minimized during preparation for market. Harvested bananas are carried to the platforms set up along the conveyor route, then lifted and hung from hooks attached to the wire. Transport speed is controlled by workers who lead the produce to the packinghouse at the bottom of the hill.

Steps to Reduce Post-harvest Breakdown

- *Harvest produce at the correct stage of development (avoid over-ripe or immature)*

  Recognizing the differences between various harvested crops will help to determine the appropriate way to handle the particular crop.

  Immature or overripe produce will store differently than mature produce. The more mature a harvest product is, the harder it is to store and the quicker it needs to be marketed. Overripe produce often breaks down rapidly and will accelerate ripening of adjacent produce.

  Immature produce may need to be handled differently to ensure marketable quality. Some crops will ripen after harvest, provided that they are physiologically mature, while others will stay at the stage and condition they are harvested at.

- *Minimize wounding or damage*

  Minimizing harvest damage can improve storability of most crops. Wounds are an open door for the entry of post-harvest pathogens and can increase moisture loss dramatically.

Curing can improve storability of onions and potatoes, as this can allow some level of wound healing and will tighten up skins. These same crops will benefit from a top-kill procedure or lifting, as this improves skin set and preparation for harvest.

Careful and minimal handling of most crops will reduce the level of damage and improve the quality. Some crops should be packaged in the field to reduce handling.

- *Consider the temperature of the harvested product*

  Harvested produce continues to be physiologically active after removal from the plant. For most produce, removing field heat will slow aging and decrease metabolic activity and respiration. Harvesting the crop in cool conditions will reduce the amount of heat that needs to be removed and will reduce cold storage costs.

  Heat removal can be accomplished in a number of ways, including packing in ice slurries or using forced air, vacuum cooling or room cooling. The choice of cooling method will depend on the type of produce, as some crops are sensitive to free or surface moisture, some crops are stored in bulk or separately and some crops will not tolerate excessive cooling.

- *Carry out grading procedures*

  Post-harvest grading can include trimming, washing, grading for size/uniformity, etc. and will improve salability. Removing material such as damaged leaves, cull product, excessively damaged or diseased produce will improve the overall post-harvest quality. Damaged material can accelerate ripening of adjacent produce and increase the likelihood of disease development and spread.

  Washing of produce is acceptable for some crops, as this improves salability and removes dirt, pathogens, spray residues, etc. Wash water must be disinfected regularly, as contamination can take place. Note, some crops are sensitive to disinfection products (e.g. bleach).

Tips for Post-hrvest Handling

Beets and Carrots

- Some say to wash these root crops before storage. We do not agree, for two reasons: getting them wet can encourage decay, and washing them will remove much of the beneficial bacteria that occupy the thin film of soil on the roots. These bacteria actually help fight decay.
- Instead, we recommend gently removing soil clods from the roots, being careful not to use anything abrasive that may scratch the root surface.
- It is ideal to harvest in dry conditions, when the soil will easily slough off of the roots.
- Wash them as you remove them from storage for eating throughout the winter.
- For beets and carrots, it is also important to trim the tops off close to the root — leave about 1/4 inch of tops material there. Leaving any more than this will invite decay, but not leaving anything will hasten the drying out of the root.

BEETS — Cold and Moist

All varieties of beets can be kept for up to 5 months under cool, moist conditions.

- Harvest before the first hard freeze, at about 1¼–3 inch in diameter.
- Trim tops (stems and leaves) to ¼ inch in length.
- The taproot should be cut off with a sharp knife prior to storage.
- To store, pack in perforated plastic bags or in sealed containers filled with damp sand.
- Beets of all varieties will keep for 3–5 months when stored at 32° F/0° C and 90–100% humidity.

CARROTS — Cold and Moist

'Bolero' is a late-season favorite: perfect for late fall and winter harvest, maintaining good flavor, sweetness, and crunch with long-term storage.

- Harvest carrots for storage before the first hard freeze.
- Trim tops to ¼ inch length. To store, place in perforated bags, or pack in damp sand in sealed containers.
- Carrots are sensitive to ethylene gas emitted by certain fruits (such as apples), so be sure to keep them separate.
- Store at 32° F/0° C and 90–100% humidity.

Celeriac (Celery Root) — Cold and Moist

'Brilliant' celeriac: pure celery flavor with excellent storage potential.

- Harvest celeriac prior to the first hard freeze.
- Trim tops to ¼ inch in length.
- Store harvested celeriac with soil and roots intact.
- Can be placed in perforated bags or packed in damp sand in a sealed container for storage.
- Clean before selling.

Kohlrabi — Cold and Moist

Big and beautiful, the rounded bulbs of 'Kossak' kohlrabi can be stored for up to 4 months.

- Harvest storage kohlrabi while the tap root is still round, before it begins to elongate.
- Remove leaf stems and tops prior to storing.
- Can be placed in peforated bags for storage.
- Store at 32° F/0° C and 90–100% humidity.

Parsnips — Cold and Moist

'Albion' is a parsnip variety well suited to fall harvesting.

- Mow tops then broadfork or undermine, or use root crop harvester.
- Parsnips require a full season of growth, and their sweet flavor is brought on by cold weather. Harvest in the fall or leave in the ground through the winter.
- When harvesting in early spring, dig before the tops begin to re-grow for the highest quality roots.
- Storage conditions are the same as carrots — hold unwashed (or washed) in perforated bags or bins at 32° F/0° C and 95% relative humidity — but they should be handled with more care as they bruise more easily.

Potatoes — Cold and Moist

'Elba' organic potato has excellent storage potential.

- Plants are mature when foliage naturally dies back. Late-maturing varieties may need to be flail-mowed to encourage maturity prior to frost.
- Tubers should remain in the ground for at least 2 weeks after foliage has died back to allow for skin set. Do not allow tubers to freeze, as they will become watery and unusable.
- Dig tubers and allow skins to air dry for a day if rainy weather is not expected.

- Do not wash tubers or put wet tubers directly into storage.
- Place in mesh bags, crates, or vented boxes.
- Store in a dark cooler at 40° F/4.4° C and 95% relative humidity.
- Healthy tubers can store for 5 months or more under the proper conditions.

Rutabaga — Cold and Moist

Shown here: Sweet, mild 'Laurentian' rutabaga

- Harvest rutabaga when roots reach the desired size, preferably after a couple of good frosts.
- Remove tops.
- Store at 32° F/0° C and 90–100% humidity.
- Clean roots may be waxed prior to delivery at market to prevent drying, but this step is not necessary.

Sweet Potatoes — Cool and Moderately Moist

'Mahon Yam™' — when cured properly, this organic sweet potato has excellent storage potential.

- Sweet potatoes can be dug when they reach the desired size, but always harvest prior to frost, as plants and roots will be damaged.
- Clip vines at soil surface and dig tubers with fork.

- Handle tubers very carefully to avoid damaging skin; do not wash.
- Cure tubers in a warm (85° F/29° C) dark place with good ventilation and 85% relative humidity for 5–7 days.
- Place tubers in crates or vented boxes and store in dark at 60° F/16° C and 85% relative humidity. Do not allow the storage temperature to drop below 50° F/10° C, as this will chill and injure the tubers.
- Once cured, store the tubers for 3–4 weeks before selling and/or consuming for better sugar content and eating quality.
- Properly handled tubers can be stored for 7 months or more.

Turnip — Cold and Moist

'Purple Top White Globe' turnip can be held for up to 4–5 months under cold, moist conditions.

- Harvest turnips when the roots have reached the desired size.
- A light frost can enhance flavor.
- May be waxed, but this is not necessary.
- Store at 32° F/0° C and 90–100% humidity.

Alliums

Leeks — Cold and Moist

'Bandit' and 'Lexton' can be stored in the field into winter.

- Leek varieties vary in the length of time they require to reach maturity ("days to maturity" or DTM), and the length of that period is directly related to their cold tolerance:
    - *Early* leeks grow faster and are ready for harvest in late summer to early fall, but they are less cold-tolerant.
    - *Midseason* leeks are ready in fall, and can be relatively cold tolerant.
    - *Late* leeks are ready in late fall to early winter, and are the most cold tolerant. These late types can be held in the field and harvested as needed. In milder climates, plants may overwinter with some protection.
- Harvest early/summer leeks first, moving on to midseason/fall leeks, and then late/winter leeks.
- Lift plants with fork to harvest.
- Plants will store several weeks with either method below:
    - Clean plants by trimming tops, roots, and peeling outer leaves. Store in boxes at near freezing (32° F/0° C) and 95% relative humidity.
    - Trim tops and peel any necrotic leaves. Trim roots, but leave an inch or two. Store in a root cellar, plants upright in a container with roots in a moist soil/sand/peat mix. Exposing to some light will keep tops green. Fully trim roots and peel outer leaves as necessary to clean up prior to selling/use.

Onions — Cold and Dry

'Redwing,' 'Patterson' & 'Cortland' can be stored 6+ months.

- When necks are soft and tops are falling (about 50% of plot), pull plants and cure in field for 2–7 days, depending on weather. Cure in warm (80° F/27° C), dry area (barn or shaded greenhouse) if rainy weather is expected. Too much sun exposure can result in bulb sun-scald.
- Bring in to a protected area (barn or shaded greenhouse) and allow plants to finish drying. Skins should be dry; necks should be dry and tight, and should not slip when pinched.

- Trim tops about 1” from bulb and trim roots.
- Place in mesh bags, crates, or vented boxes. Store in a dark cooler that is near freezing (32° F/0° C) with 65–70% relative humidity.
- Sweet/mild onions have much higher water content, and thus will not store longer than a few weeks. Pungent storage onions typically store 4–6 months or longer.
- Sort onions on a regular basis to remove any rotten bulbs.

Shallots — Cold and Dry

Shallots — a welcome ingredient in winter recipes.

- When necks are soft and tops are falling over, pull and sun-cure at least 2–7 days, depending on weather. Move to a protected location to finish drying.
- When dry, clip off tops and roots, and store in onion bags or shallow boxes at near-freezing temperature (32° F/0° C) and 65–70% relative humidity.
- Shallots can last in storage up to 6 months.

Brassicas

Brussels Sprouts — Cold and Moist

‘Diablo’ & ‘Nautic’ can be harvested after frost.

- Harvest when heads are about 1 inch in diameter.
- Once cut, they should be stored in perforated bags at 32° F/0° C and 90–100% humidity.
- Whole stalks can also be harvested and stored for up to 1½ months.

Cabbage — Cold and Moist

'Storage No. 4' has outstanding storage potential — up until spring from a late fall harvest.

- Harvest when heads are compact and firm.
- Store with a few of the outside wrapper leaves at 32° F/0° C and 90–100% humidity.
- Clean before selling.

Cucurbits

Specialty/Ornamental Pumpkins — Cool and Dry

'Musque de Provence' — a classic keeper pumpkin.

- Ornamental pumpkins are ready for harvest when color is fully developed.
- Fruits may be left in the field after reaching maturity, but overexposure to sun at this point will reduce fruit and stem (handle) color quality.
- Fruits can tolerate 1–2 light frosts prior to harvest.

- To harvest, clip handles close to vine. Avoid picking up pumpkins by handles, and take care to not damage skin/rind.
- Sun cure in the field for 5–7 days, or cure indoors by keeping fruits at 80–85° F/27–29° C with good air ventilation for 5–7 days.
- White varieties should be brought in out of direct sunlight once foliage starts to die back; cure inside and keep out of sun to avoid yellowing.
- Some growers prefer to wash fruits with a mild bleach solution after curing.
- Store at 50–60° F/10–15° C, with 50–70% relative humidity and good ventilation.

Winter Squash and Pie Pumpkins — Cool and Dry

'Waltham Butternut' has storage potential of up to 6 months.

- Winter squash and pie pumpkins are generally mature about 50–55 days after fruit set, and should be harvested before hard frost.
- Cut fruits from vines and handle carefully.
- Sun cure in the field for 5–7 days or cure indoors by keeping squash at 80–85° F/27–29° C with good air ventilation for 5–7 days.
- Store at 50–60° F/10–15° C, with 50–70% relative humidity and good ventilation. Repeated exposure to temperatures below 50° F/10° C may cause chilling damage.
- Storage potential and timing of best eating quality vary by type.

Harvesting of Crops

Harvesting is the act of removing a crop from where it was growing and moving it to a more secure location for processing, consumption, or storage. Some root crops and tree fruit can be left in the field or orchard and harvested as needed, but most crops reach a period of maximum quality—that is, they ripen or mature—and will deteriorate if left exposed to the elements. While the major factor determining the time of harvest is the maturity of the crop, other factors such as weather, availability of harvest equipment, pickers, packing and storage facilities, and transport are important considerations.

Economic and marketing issues are often even more important than considerations of maturity in deciding when to harvest a commodity. Before the crop can be harvested, the grower must be sure that there is a demand for the crop and that the price is sufficient to make harvesting the crop profitable. If the price is less than adequate to cover the costs of production, harvesting, and marketing, growers are faced with the difficult decision whether to harvest and store the crop, to wait for a better market, or to cut their losses and leave the crop in the field.

The Process of Harvesting

Harvesting can be separated into three steps. The plant part of interest must be identified, detached from the rest of the plant, and then collected in a container suitable for transport from the field. The harvesting of all the major agronomic crops (grains of cereals and legumes) has been mechanized. The resistance of dried cereal and legume seeds (for example, corn, rice, wheat, and soybeans) to physical damage allows the first and second steps to be combined in a threshing machine or combine that separates the seeds from the rest of the harvested plant. The grain (seeds) is then loaded in bulk containers and transported to silos for additional cleaning, grading, fumigation, and temporary storage.

In contrast to the dry grains and legumes, most horticultural crops (fleshy fruits such as apples and tomatoes, ornamentals, and vegetables) are hand harvested for the fresh market. Some mechanical harvesting is done, but the damage incurred is usually so severe that the fruits and vegetables are only fit for processing. Some fruits (for example, apple, pear, and tomato) evolved to ensure seed distribution by enticing animals to eat the attractive (in appearance, aroma, and flavor) fruit. Therefore, they are almost always attached to the plant by a specialized structure that contains an "abscission zone" which permits the ripe fruit to be easily separated from the tree. Other fruit, such as bananas, citrus, peppers, and all vegetables (for example, cabbage, broccoli, carrots, lettuce, and squash) do not develop an abscission zone and must be cut from the plant.

Only human beings have the unique combination of eyes, brain, and hands that permits the rapid identification and harvest of delicate and perishable materials with minimal loss and bruising. Some mechanical aids such as cherry pickers, ladders, picking bags or baskets, stem clippers, and wheelbarrows are used by harvesters to make their work easier, faster, and safer. Harvesters can be trained to select only those fruits or vegetables that are of the correct maturity, size, or shape, thus greatly reducing the amount of material that must be removed on the sorting and grading line in the packing shed. In fact, many vegetables and berries are harvested directly into retail containers without further sizing or grading. Most other horticultural crops are harvested into field bins that are taken to packing sheds where the commodities are cleaned, sorted, graded, inspected, packed, cooled, and stored before being transported to regional markets.

Harvesting rice consists of the basic operations which can be done in individual steps or in combination using a combine harvester. These include:

- Reaping - cutting the mature panicles and straw above ground.
- Threshing - separating the paddy grain from the rest of cut crop.
- Cleaning - removing immature, unfilled, non-grain materials.
- Hauling - moving the cut crop to the threshing location.

- Field drying - leaving the cut crop in the field and exposing it to the sun for drying *(optional).*
- Stacking/piling - temporarily storing the harvested crop in stacks or piles *(optional).*
- Bagging - putting the threshed grain in bags for transport and storage.

Traditional harvesting activities such as field drying and stacking/piling are not recommended because they can lead to rapid quality deterioration and increased harvest losses.

Besides these, a variety of other activities can be included in harvesting such as gathering, reaping (gathering standing grain by cutting), bundling, and various forms of transporting the crop and grain.

Crops may be harvested at different stages of fruit maturity depending on the produce and factors such as:

- The market destination.
- The purpose of the produce i.e. is it to be used as fresh or processed produce.
- The time it takes for the produce to reach the consumer.
    - If the produce is aimed at the local market the fruit may be harvested when ripe or are about to ripen.
    - If the produce is destined for the export market the fruit can be harvested when fruit is mature, but not ripe yet.
- The external appearance of the produce is less important when the fruit is meant for processing. There remains quality standards such as:
    - Water, sugar and acid content of the fruit.

For vegetables, seed packages give an indication of "days to maturity". This indication can also be found in seed catalogues. Because the ripening of a crop is largely dependent on environmental conditions the time indication of "days to maturity" should only be used as a guide.

The best way of determining the readiness of the crop for harvest is to check the garden frequently for ripe produce during close to the expected harvest time.

When harvesting, avoid bruising or damaging the vegetables as this will lead to decay of the produce.

Vegetables continue to carry on life processes even after harvesting. It is thus important that we retard these processes in mature products. With the exception of ripening storage of vegetables does not improve the quality of the produce. A vegetable must possess high quality characteristics and be cultivated properly for it to be acceptable.

It is not easy to determine the point at which specific crops reach their peak quality. It is essential that one keeps accurate and complete records of the varieties that you use, when they were planted etc. This will help you gauge the ideal time of harvest.

Some crops are more easily damaged during harvesting than others. However bruises and cuts must be avoided during handling with all crops. All produce showing decay or rot must be removed and disposed of.

Washing certain vegetables in cold running water immediately after harvesting removes soil, dust or other contaminants and helps lower the temperature. By adding a suitable sanitizer, microbial contaminants may be removed.

When storing vegetables the following requirements need to be considered:

- Temperature

  The conversion of sugar to starch is critical in products such as sweet corn and peas. It is necessary to cool these products immediately after harvest to minimize this conversion. If possible, harvest vegetables early in the morning or right before you intend to use them.

- Moisture

  Proper humidity varies with different commodities. In general however leafy vegetables need a high humidity (95 %) whereas onions should be stored in a drier atmosphere (65 to 70.5 relative humidity).

- Ventilation

  Wilting and tissue breakdown are minimized if the storage area is well ventilated.

In preparation for harvesting, an assessment is done of the tools and equipment required. This is done taking into account the size of the farm, the crop and cultivar, the period during which the harvesting must be completed, and the capacity of storing facilities, where if applicable.

Once this information has been gather a decision can be made as to the number and size of picking teams, which in turn determines the equipment requirements.

In the case of mechanical harvesting such assessments will determine when a harvester should be booked, if the farmer does not own a harvester, and the number of harvesters that may be required. In addition the numbers of back-up equipment such as tractors and trailers used for transporting harvested goods can be decided upon.

Harvesting Methods

The crop type and expected harvest volume will determine the harvesting method that is to be used. Certain crops are harvested by hand, while others are harvested either manually or mechanically. Hand harvesting usually provides a better quality product than mechanical harvesting. Harvesting by hand is however more time consuming and expensive than mechanical harvesting and is not always practical or economical feasible.

The harvesting method to be used is also affected by the area (size of the field) to be harvested. A farmer might easily be able to harvest half a hectare of maize by hand, but it is not possible hundreds or even thousands of hectares are to be harvested.

In addition the market may often dictate the harvesting method. If for example tomatoes are to be sold on the fresh produce market, harvesting will be by hand in order to ensure that the crop is not damaged. If, however, the fruits are destined to be processed a certain degree of damage can be tolerated at harvest.

Crops that are exported as fresh produce, including most of the tree fruit and vegetables, are traditionally harvested by hand. The quality of fruit and vegetables play an important role in their marketing and thus the amount of damage should be minimised. Physical damage also leads to shortened produce shelve life.

Crops such as tobacco, and cotton are examples of industrial crops, which are harvested by hand. In the case of tobacco this is because the leaves mature from the bottom of the plant up. The time lapse between ripening of the bottom to the top leaves is such that the plant cannot be harvested at once.

Hand picked cotton is of a higher quality than machine picked cotton, because hand picked cotton can be picked sooner and cleaner than with a harvester.

Grain (wheat and maize), oil (sunflower and canola) and protein (soybean and groundnut) crops are seldom harvested by hand. These crops are harvested using specialized combine harvesters.

There are instances where a combination of hand and machine picking is practised. An example is where onions or potatoes are lifted out of the soil using machines.

The bulbs or tubers are then picked up and sorted by hand.

Harvesting Equipment

Harvesting equipment may include a wide range of different instruments and containers.

Ladders are used especially in tree crops to reach fruit high in the tree. The ladders should be kept clean and in good condition, ensuring that pickers do not fall and injure themselves.

Picking bags are used especially in orchard, as they are easier to handle that crates or baskets. The bag is slung across the shoulder and fruit placed carefully into the bag after being picked. The bags must be kept clean and not be thrown on the ground or walked on. This will ensure that no sand and grit gets into the bag, which can damage the produce.

Where bins are used the same principals apply. The crates must be keep clean and should not be sat on.

The bins into which produce is placed must be handle with care so-as they do not break or splinter. Splinters in the bins will perforate the harvested produce.

Picking knives and shears. In crops such as lettuce and cabbage, harvesting knives are usually used. The knife is placed between the first and second leaves and a single clean cut removed the head from the stem. The knives must be sharp at all times; otherwise it will not give a clean cut. The knife must be kept clean at all times. In between cutting, the knife must be dipped in a sanitising solution. This will ensure that potential post-harvest diseases are carried from one head to the next.

In the case of fruit harvesting, fruit are harvested either by hand or some for of harvesting sears or cutter. These are used to cut the fruit stem from the branch. In this way you ensure that the fruit stem is not torn from the fruit. If the fruit tears, it will enable entry of disease and lead to rotting of the fruit. It is important that the fruit stem is not left too long as it could puncture the accompanying fruit in the bags on bins.

All equipment must be cleaned and stored in a designated storage area for the harvesting equipment. The equipment must be kept dry and should not be kept on a dirty floor. Harvesting equipment must not be stored in close vicinity to pesticides, pesticide application equipment or fertilisers. The equipment could become contaminated and thereby contaminate the harvested produce.

All tools and equipment used in the harvesting process must be properly cleaned and sterilized before and after use as well as at regular intervals during harvesting.

Workers must also wear clean and suitable protective clothing.

- Pruning shear

  The most common tools used for harvesting fruit are steel pruning shears. They are used to cut the stem of the fruit as close to the button, or calyx, as possible without injuring the shoulder of the fruit. Some vegetable fruits, like tomatoes, can easily be picked by a simple twist of the hand, while pruning shears are used on others like peppers.

  In some crops, specialised shears and snips are used for harvesting. Pruning shears fit comfortably into the palm of the hand, have a rounded point and curved blade to avoid damaging the fruit. A spring keeps the blade in an open position and ensures fast recovery for efficient picking.

  In preparation for harvesting, all pruning shears must be checked to ensure that the spring action is working well, and that the blades are sharp and closely aligned. This should also be checked on a regular basis during picking. Pruning shears must also be cleaned and sterilised and prior to harvesting and regularly during picking.

- Picking containers

  A number of receptacles are used to collect picked fruit, but the most common is a picking bag. Picking bags are generally made from canvas or polyvinyl. They have a single strap and are slung over the shoulder of the picker. As the picker picks the fruit, it is placed inside the bag until it is full. It is then taken to a general collection point in the orchard where the bags are emptied into bulk bins or picking trailers.

  A number of different bag designs, are available and used in various industries.

  In preparation for harvesting, all picking bags and containers must be checked for holes and/or tears. The containers must be cleaned before harvesting is initiated. When cotton is hand picked, the cotton is placed in picking bags before being emptied into metal baskets. Unlike for tree crops the bags are non-woven bags (material or polyvinyl) as the strands of these bags contaminate the cotton fibre. Cotton picking bags most often used are empty plastic fertilizer bags. Because these bags are not porous and the cotton will still contain moisture, picked cotton should not be left in these bags overnight as the cotton may rot.

Fruits and vegetables, which damage easily, are picked into picking crates taken to the cold rooms or packing sheds as soon as possible. These crates are usually made from durable plastic and are well ventilated. They are also designed to be stacked.

Clean, dry and sterilized crates should be used to prevent disease contamination. Crates, which are damaged in such a way that bruising or puncturing of the crops could occur should be discarded.

- Harvesting Knives

  Knives are used in harvesting leafy crops such as cabbage heads, Swiss chard leaves and lettuce. A good quality knife with a strong blade should be supplied to the harvesters. The blade should be sharp enough to facilitate easy cutting. A blunt knife will not only reduce the harvest speed, but can cause unnecessary damage to the produce. Knives should be regularly cleaned and disinfected whilst harvesting to prevent the spread of post harvest decay organisms.

- Ladders

  Ladders are used during harvesting of tree crops depending on the height of trees. Ladders are placed alongside trees, or rested against the canopy in the case of larger trees.

  Ladders vary from home-made, single-sided, two-legged ladders made from locally grown wood to factory-manufactured, two- or three-legged ladders generally made from aluminium or wood.

In preparation for harvesting, ladders must be checked for sharp edges, splinters, loose components and dirt to ensure fruit is not damaged when the ladders are placed alongside the trees or against the canopy. The ladders should also be checked for defects that may compromise the safety of the harvesters.

- Mechanical harvesters

  Field crops such as grains, vegetables produced for industrial purposes, potatoes, onions, oil crops, protein crops and cotton are harvested mechanically. The different types of harvesters are specially designed to for a specific crop or group of crops.

  Wheat, maize, sunflower and soybean can be harvested with combine harvesters. These types of harvesters harvest the whole plant. This usually involves fast moving rollers and or blades.

  The harvester then removes kernels from the ears or the seeds from the heads or pods in a process known as threshing. It is a complex mechanical action that results in clean seed being dropped into a storage compartment. The remaining plant residue is released (stalks, straw, pods, leaves, etc.) back onto the field. As soon as the storage compartment is full, it is offloaded into bulk trailers and transported to silos where it is stored.

Servicing the harvesters should be done before the start of the season to prevent hold-ups due to breakage during the harvesting process. During harvesting, any blockages must be dealt with as soon as it occurs, as permanent damage may result. The harvesters should be cleaned at the end of each day to prevent any hold-ups the following day. While cleaning one can also easily pickup any problems such as broken blades or other damaged parts etc.

Cotton harvesters have rotating comb-like spindles, which comb the cotton out of the open cotton bolls. At the back of the harvester, metal baskets, are attached into which the cotton is blown. When a basket is full, it is emptied into another basket in which it is then transported to the gin. Harvesters should be serviced before the start of the season.

Onions, potatoes and groundnuts grow below the soil surface. Commercially these crops are harvested by lifting the bulbs, tubers or pods from the soil before. Once lifting has been completed the crop is picked up by hand, sorted and packed. Lifting is done using specialized implements containing a blade, which is attached to the back of a tractor. The "blade" is then inserted into the soil at one end of a row and drawn just below the soil surface (at a depth of ± 15 – 20 cm).

The next step can be manual or mechanical. In both cases it involves the pulling of the plants from the soil and removing the tubers etc. from the plant. Potato tubers can be placed in bulk trailers or bulk bins on trailers and transported to the pack houses. In the case of onions the bulbs are first left on the field for at least a week to cure before the dry leaves are cut (with sharp knives) from the bulbs and the bulbs placed in onion bags, ready for the market. Peanut plants are also left on the field for a week or two in heaps or in windrows to cure. Then only will the plants will be threshed and the pods placed in woven material bags or bulk trailers and transported to where the peanuts will be processed.

- Garden fork

  In the case of smaller production areas, where mechanical harvesters cannot be used, a fork can be used instead for lifting produce. The fork is inserted into the soil at an angle, close to the crop plant and then the plant and soil are lifted. Take care not to push one of the fork teeth into a tuber, bulb etc.

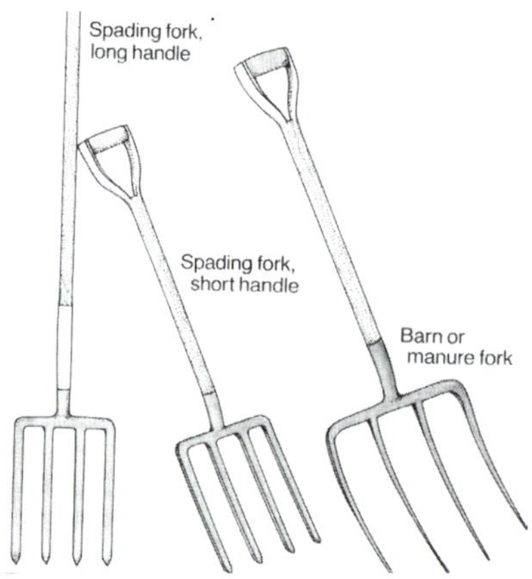

- Bulk bins and picking trailers

  Fruit are transported from orchards in bulk picking trailers with a capacity of 2t to 3t or in wooden or plastic bulk bins with a capacity of 350 kg. In small operations, crates with a capacity of 15 kg to 20 kg capacity are used.

Bulk bins are normally supplied by pack houses, or rented from an equipment supplier. In preparation for harvesting, bulk bins are checked for sharp edges and splinters to avoid damage to fruit. They are also cleaned and sterilised to prevent contamination and decay. The supplier of the bulk bins normally does this.

All moving parts of picking trailers must be serviced before the harvest starts. Also check the tyres to make sure they are in good order. If trailers have to be used on national roads, they must be roadworthy. The fruit carrying area of the bin must be checked for dirt, sharp edges, latch functioning, etc.

The metal baskets in which the cotton-picking bags are emptied and in which the cotton can be stored, should be dry and clean before the seed cotton is placed therein. Because it is made from metal, any rust should be removed and treated.

- Cleaning, sterilizing and maintaining harvest tools and equipment

  Cleaning of harvesting tools refers to both removal of dirt and plant material becomes adhered as well as to the removal of microscopic particles including disease organisms.

  The harvested produce must not come into contact with post-harvest decaying organisms, as it will reduce the shelf life and quality of produce.

  Removal of dirt, plant material etc. from tools and equipment prolongs the lifespan of the tools.

Harvesting Processes

Before harvesting can commence, a number of decisions must be taken and preparations must be made.

The harvesting process can be divided into sections. If for instance a reasonable percentage of the fruit in a certain orchard will not be fit for export because of external damage, a decision must be taken to pick the fruit that is more likely to be fit for export separately. Note that pickers who are entrusted with pre-selection in the orchard/field must be well trained and closely monitored to ensure that export fruit is not picked by mistake.

Based on the maturity indexing data and the crop estimate, a decision has to be taken on when picking will start, and on the picking period.

A decision can also now be taken on the number of pickers and picking teams that will be required. The necessary workers are recruited and trained, if necessary prior to the harvest.

In preparation for harvesting, a general inspection of the equipment is required to ensure that maintenance has been adequately seen to. Pruning shears, picking bags, ladders, bins or trailers, and tractors are inspected to ensure that they are in the proper working condition. All equipment that may come into contact with fruit must also be cleaned and sterilised to limit waste and decay.

Mechanical harvesters are very expensive and harvesting is often contracted out. It is therefore essential for a farmer to book the harvesters well in advance to prevent yield and quality losses due to too late harvesting. Contractors usually prefer to harvest all the fields in a particular production area before moving on to the next production area and will seldom return to an area in the same season. Therefore the planting time of the crops should also be seen as part of the harvest preparation, because if you plant earlier or later than the other farmers in the area, your crops development will not be in synchronization with the rest of the fields in the area and that can lead problems come harvest time. If a farmer, however, has his/her own harvesters, the latter is not of importance. As with contract harvesting, make sure to enquire and plan for contract labour, transport of the produce to markets, gins, silos etc., if it has to be contracted out.

Harvesting Methods

- Fruit picking methods
  - Snap picking;
  - Clipping.

Snap picking is a method whereby fruit is twisted off sharply by hand without using an implement. Clipping makes use of specifically designed pruning shears that are used to cut the stem of the fruit. Clipping is generally recommended in most situations, but there are cases where snap picking, which is faster, is appropriate.

In the case of clipping, purpose-built pruning shears are used to cut the stem as close to the button, or calyx, as possible without injuring the shoulder of the fruit. Long stems of between 1mm and 4mm, which result from inefficient cutting, are a frequent problem in harvesting and can lead to injuries to fruit when they are put into the picking bags, trailers or bins. Injuries of this nature can cause the fruit to decay at a later stage. If fruit is handled roughly or if the shoulders are cut by inexperienced pickers or pickers who are not taking enough time, it also increases the risk of fruit decay.

Snap picking is not recommended for produce such as export citrus fruit for a number of reasons. The calyx buttons are pulled out easily, oleocellosis develops because of the pressure on fruit as it is pulled and twisted, and stems are sometimes not cleanly broken off. It is however appropriate in certain situations, such as when a team is sent in to select and pick non-export fruit.

Snap picking is often done in tomatoes and other vegetable fruit crops, while the use of clipping is often done to improve the quality of the produce less bruising and other physical damage associated with snap picking occurs.

- Harvesting of leafy vegetables

  In most cases the cabbage and lettuce heads are cut from the root at the base of the head and as close as possible to the soil surface, using a knife. Where individual leaves are harvested such as Swiss chard (spinach) or lettuce, the fingers are placed close to the base of the petioles and while firmly keeping hold of the petiole the leaf will be removed from the plant in the same way as the snap pick procedure followed in tree fruit. This may, however, cause tearing of the petiole tissue and knifes could also be used for a cleaner cut.

- Harvesting of grain and seed crops

  If small areas are to be harvested, maize ears can be harvested with the snap picking, while pruning shears can be used cut the heads of the sunflower stem. The ears can then be taken to the shed where the ear leaves have to be removed to gain access to the kernels. By simply rubbing the hand palm firmly over the ear, the kernels can be removed. An alternative is to rub two ears together. When soybeans are harvested it would be easier to cut the stem with pruning shears or any other sharp tool close to the soil surface, taking the stems to the shed and manually removing the pods from the stem there. The seeds can be removed by pressing hard on the sides of the pods to force it open.

Guidelines to Harvesting Crops

The section below summarises harvesting guidelines for selected crops:

- Citrus

  Once picking starts, a few basics guidelines should be adhered to, as set out below:

  - Picking bags are suspended from the side of the pickers and not in front of them. This prevents damage to the fruit as the picker leans against a ladder or pushes into branches in the canopy.
  - Pickers must walk with full picking bags and not run, as running will bounce and chafe the fruit, leading to the development of oleocellosis.
  - When picking bags are emptied they should be placed close to or on existing fruit lying in the bulk bins or picking trailers before being emptied. This will also limit injury.

- Once a bag has been emptied, it must be opened and shaken to remove loose twigs, leaves and sand that might have collected during the picking process.
- Fruit that were dropped during the picking process must not be picked up off the ground and put with export fruit. Similarly, fruit lying on the ground and in contact with the soil should not be picked up and mixed with export fruit.

When varieties that are susceptible to oleocellosis are harvested, filling of bulk bins and trailers should be limited. The extent of the limitation will depend on the specific condition, but it would generally never be approximately 50% of the volume of the bin/trailer. As an added precaution, rubber or cardboard sheeting can be used to line bins providing protection.

- Soft fruits such as tomatoes

  Ensure that clear instructions are given on which colour fruits should be harvested and which should be left for the next picking. Follow the site procedures where they are available. Provide examples of some poor quality fruits and what should be done if such fruit are encountered. These fruits bruise easily and should not be thrown into the picking crates from a distance

- Ears, heads and pods - dry

  As these are already dry, care should be taken to prevent the seeds from falling out of the head or pod before it reaches the threshing area. At this stage some diseases could have attacked the ears, heads or pods. The pickers should know how to identify these and should separate diseased pods and cobs.

- Vegetable Crops

  Vegetables should be harvested during the cool part of the morning and should be stored as soon as possible. If the produce is destined for the market, ensure that the storage conditions are correct for the produce. If they are destined for processing, they should be cooled and stored under refrigeration to preserve flavour and quality. You need to ensure that you know the basic harvest practice for each crop you are dealing with. What follows are some examples of harvest practices for different crops:

  - Beans – green - Start harvesting before seeds develop in the pod – when the green pods are about the diameter of a pencil. To determine harvest readiness bend them in the middle, if they snap easily, they can be harvested.
  - Beetroot - Harvest when they are about 5 – 8 cm in diameter. If the market is for smaller beets they can be harvested at about 2 – 3 cm diameter. The leaves should be about 10 – 15 cm long. The beet tops can also be eaten as greens.
  - Broccoli - Harvest the dark green compact clusters or heads when they are about 15 cm in diameter. The buds must be tight and should be harvested before any yellow flower develops. As you harvest the heads, smaller side shoots will develop providing an almost continuous harvest.
  - Cabbage - Harvest cabbage when the heads feel hard and solid. If left to late the heads will crack and split. Cut the heads from the stem just below the point of attachment to the stem. Ensure that you use a clean, sharp suitable knife.Always clean the knife using a dip mixture of suitable sanitizer in water. One can harvest the sprouts that develop as a later crop.

- Cucumbers - Harvest when the fruits are deep green and before a yellow colour develops. The cucumber fruit should be 5 – 8 cm long for sweet pickles; 13 – 16 cm for dill pickles, and 16 – 20 cm for slicing pickles. Harvest 4 to 5 times per week to encourage continuous production. Mature cucumbers left on the vine will stop the production of the entire plant.
- Lettuce - For non-heading types harvest the older, outer leaves from leaf lettuce when they are approximately 12 cm long. For heading types harvest when the heads are moderately firm but well before seed stalks form.
- Onions - Harvest when the tops fall over and begin to turn yellow. Dig out the onions and place them to dry out in the open sun for a few days. The onions are sufficiently dry when skin is toughened. Remove the dried soil by brushing the onions lightly. Cut the stem, leaving 4 to 5 cm and store in a net bag in a cool, dry place.

• Tree crops - general

- Do not shake the branches of the trees to make the fruits fall down as this will lead to bruising.

- Use ladders where possible.
- When harvesting for processing, a harvesting stick can be used where fruit are to high to reach but allow the fruits drop onto a soft surface, such as a net.

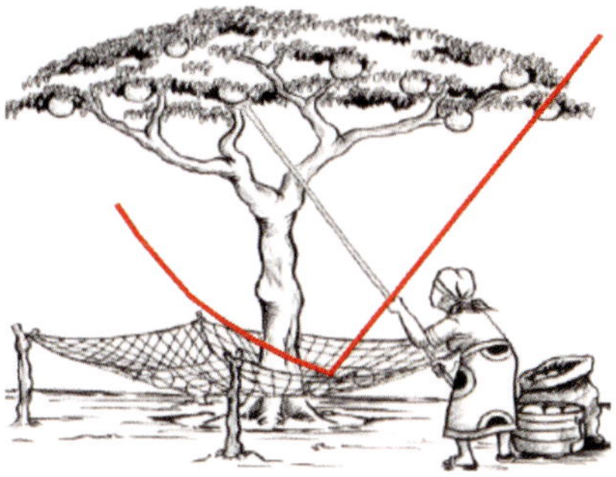

- When mango and avocados are to be exported or transported for extended periods, leave a portion of the stalk on the fruit.
- Do not throw fruits out during harvest.

- When harvesting heavy banana bunches use two people per bunch.
- Harvest in the early morning or late afternoon.
- Do the first grading according to market destination during harvest.
- Ensure that the latex from mango, banana and paw paw does not come into contact with the skin of the fruit. If this happens wash the fruit in clean water.
- Take care of your eyes, if harvested fruits contain latex juice.
- Place your fruits on a well-aerated surface or in containers that are not made of rough surfaces.
- Handle the fruit as little as possible.
- Pack the fruits as soon as possible into the final transport box or container.
- Export fruit may have to be cooled.
- The basic principles of orchard hygiene are:
    1. In order to avoid mould infections and rind damage a number of standard orchard hygiene practices could also be implemented.
    2. Prune trees of dead branches twigs.
    3. Prune the trees to allow light and air movement through trees.
    4. Harvest cultivars that are susceptible to splitting early.
    5. Never place fallen fruit into bins together with fresh fruit.

Harvest Timing

In some cases fruit need to be tested for parameters such as sugar contents or acidity etc. prior to harvest. In order to generate accurate analyses results however, one needs to take a sample of fruit or produce that represents the status of the fruit of the field or orchard. To this end the samples to be taken must be taken randomly from the field. To do this, divide the field into equal parts or sampling units. If a tree crop is involved each tree could be viewed as a sampling unit. Number the sampling units starting from one. Then using a random number generator (on a calculator) or random number tables, select 5 sampling units that to be sampled. Once selected go to those specific sampling units and sample the area. In the case of tree crop divide the tree into 4 quadrants. Pick a fruit from each quadrant representing and from each of the lower middle and upper part of each quadrant. Also sample from the outer reaches of the branches and the inner (closer to the stem).

- Maturity Indexing and Sampling

  Before a crop can be harvested, it must be determined whether the crop is ready to be harvested. To this end maturity indexing is done for a crop.

  Ready for harvesting does not necessarily mean that the fruit is mature or ripe. Different crops are harvested at different stages of development. Baby vegetables for example are harvested at a young stage mostly before the crop is mature. In vegetables such as carrots, beetroot and lettuce the size requirement for the crop may determine when it will be harvested.

Fruit crops and fruit bearing vegetables can be divided into two groups with regards to harvest. In the one group the fruit is harvested at a fairly green stage because full ripening will only occur post harvest. This often helps to prolong the shelf life of the produce. Examples of these crops include avocados, mangoes, bananas and tomatoes. Other fruit crops such as grapes, litchi's and strawberries will not further ripen or improve quality after harvesting. This means these types of crops must be harvested when mature.

Maturity indexing information is used to plan harvesting, by determining whether fruit should be picked earlier or later than the historically recorded picking dates, and adjust orchard practices where needed.

Maturity indexing also provides information on how the internal physiological development of the fruit is progressing relative to other physiological factors, such as the colour and size. Equally importantly it allows the manager to assess how the fruit timing, or readiness for picking, and quality relates to the specific requirements and opportunities in the market.

The basic parameters that are assessed and monitored during maturity indexing are internal quality parameters and fruit colour, as defined by colour plates (Available from the different growers' associations retailers and exporters).

There are four stages of ripening for tomatoes, pale green, early light red, light red (shoulders of fruit not full red yet) and a full red colour. Despite the outside colour, the flesh is already ripe and harvesting at any of these stages will produce an acceptable product. Some consumers prefer pale green to early light red fruits if they go camping for long periods of time, as the fruits will turn full red over time and the chances of rotting will be minimal. However, a housewife who is preparing a meal for her family would prefer a light red to full red coloured tomato.

Maturity indexing is a common practise in tree crops like citrus and mango and prescribed norms are available. In vegetable crops like tomatoes, colour changes can be used to classify maturity, but no fixed norms have been developed.

Many fruit are harvested when a desired colouring has developed. The colour development is used as an indication of internal ripeness.

In grain and other field crops, maturity is usually determined by the moisture content of the produce. For each of these crops there are set norms in terms of moisture content. Table provides some of the norms as to the relation between moisture and harvest ability. When the crop is too dry, conditions in the silos can become instable and explosions may occur. If the moisture content is too high, problems with rotting and mould develop.

The moisture content norms at harvest for certain field crops,

Crop	Moisture content (%)
Maize kernels	± 13%
Sunflower seeds	10 to 13%
Seed Cotton	±12% (for machine harvest)
Soybean pods	10 to 13%
Groundnut seeds	± 10%

- Sampling techniques

  In order to conduct maturity indexing, the crops must be sampled and then a specific test is performed on the sample.

- Fruit sampling – example citrus and tomatoes

  Sampling for maturity indexing are generally initiated two months before the historically initiation of the harvesting period of a cultivar.

  It is important that the correct fruit is sampled on a plant. The first principle of sampling is that the sample must be representative of the bulk of the medium that is being sampled. A sample is not a piece or a section of a whole, but rather the whole reduced to a manageable volume.

  Although a fruit sample must be taken randomly, the focus should be on fruit that represent the development stage of the bulk of the crop. The focus should also be on the fruit that is likely to be harvested in the earlier part of the season, referred to as the first pick. Unusually large or small fruit would not be included in the sample.

  The sample is placed in a bag and the details of the field/orchard, the date of sampling is written on a label and sealed. Do not write on the outside of the bag, as the writing will wipe off.

- Sampling Field crops

  Using maize as an example the following applies for sampling field crops. The maize kernel goes through a number of developmental stages before maturing. One of these is the soft and hard dough stage. Pressing the kernel with a fingernail identifies the soft dough stage. If it releases a white milky substance the kernel is at the soft dough stage. This is the ideal stage for harvesting of green cob corn. If the same test is preformed and the kernel is still soft, but no milk flows from it, the hard dough stages has been reached, and at this stage the cobs are no longer suitable for cob corn. After this stage the kernels keep on losing moisture and become harder and harder. Physiological ripeness is reached when a black abscission layer develops between the ear and the individual kernel. At this stage the kernels can easily be removed from the cob by simply rubbing two ears over each other. The kernels will however still be too wet and harvesting cannot commence.

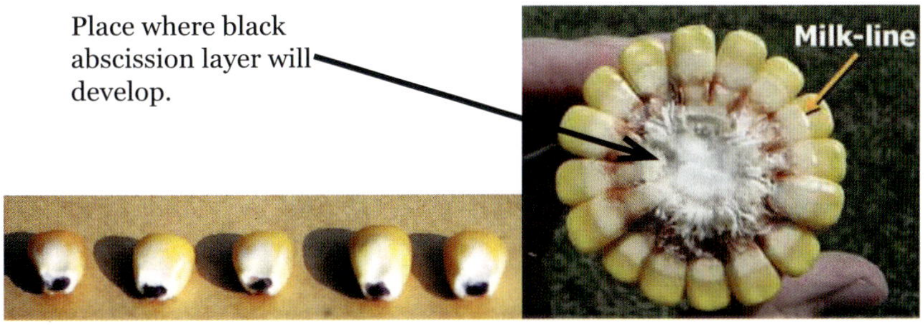

The abscission layer forming around individual maize kernels (left). On the right is an immature ear on which the arrow indicates the place where the black layer will develop.

In order to determine if the correct moisture content has been reached the following steps can be followed:

- Walk in a zigzag fashion through the maize field and pick ears from different plants spread out over the whole field.
- One should harvest enough ears to generate 5 kg of seed.
- Do not pick ears from the outside rows, as these plants are not representative of the field - Remove the leaves and thresh them.
- Place the seeds in a container and mix the seed of the different plants thoroughly.
- Now take a representative sample by taking hands-full of seeds from different places and at different depths out of the container. The sample should be at least 1 kg in mass.
- If there are no instruments available on the farm to determine the moisture content, the sample can be taken to a co-op, which handle grain crops.
- Determine the moisture content.
- If the moisture content is 15% or above, wait at least another week and repeat the procedure. Repeat this until the desired moisture content of 13 % or less has been reached.

Similar procedures can be followed for other grain crops. Talk to farmers in the area to learn from them how it should be done.

- Analysis of Fruit Samples

  The procedures set out in this topic are applicable to citrus fruits. Similar tests are also conducted on tomatoes for the tomato processing industries where the brix, acidity and juice percentage is measured. Most other vegetable fruits and leafy vegetables will not be analysed chemically, while grain, oil etc. crops will only be monitored for moisture content. In the laboratory, details relating to the sample are noted, including the producer, orchard number, and fruit size and colour. The fruit is then weighed, cut, its juice is extracted and the following parameters are measured:

  - Juice percentage;
  - Brix, which relates to the total soluble solids or sugar content;
  - Acidity.

The ratio of Brix to acidity is a further parameter used by the industry to determine acceptability of taste for different markets.

The norms for different varieties are set at different levels. In the case of lemons, the only deciding factor for export is juice percentage, as neither Brix nor acidity levels are generally used for table fruit. In the case of niche cultivars, higher standards, meaning higher Brix and ratios, have been set to place the fruit in a particular marketing bracket. That standard has to be maintained to assure the producer of the potentially higher return from that market.

- Recording Maturity Indexing Information

  Once the internal quality results are received from the laboratory, the data is assessed relative to the management plan on the farm.

  The maturity indexing information is stored in a file and presented as a graph to develop a picture of how the fruit is maturing. If orchard practices need to be changed to manipulate fruit development, maturity indexing is a key factor in guiding this.

Safety Wear Appropriate to harvesting Crops and Handling the Harvested Crop

- Working with machinery

  Operators should not touch any machine unless he/she is trained to conduct maintenance on the machinery. When machinery is involved,it is best to wear overalls, protective gloves and closed end shoes (not barefoot or sandals). All parts of the body, except the face is covered, reducing the risk of injury due to stray rocks or even plant debris during the harvest process.

  Wear protective eyewear. Fine dust, soil and plant particles are always present during mechanical harvesting of dry products such as maize and soybean, and can cause severe eye damage. Due to the dust and fine particles surrounding the harvester at work, dust masks should be used to prevent these from being inhaled. Do not wear loose clothes that may be caught in moving parts of machinery.

- Safety during hand harvesting

  Wear overalls and closed shoes. Although no machinery is involved, grass leaf blades can cut exposed skin. Some crops such as tomato contain chemicals that may irritate the skin. Unprotected feet may be injured if harvesting bins are dropped on the foot. If pruning shears, knifes or other tools with blades are used during the harvesting process, wear protective gloves.

- Post Harvest

  When machinery is involved, it is best to wear overalls, protective gloves and closed shoes. Post harvest treatment of the produce may involve the use of chemicals. In such situations ensure that the correct Personal Protective Equipment is used.

- Personal hygiene and health

  To prevent contamination of produce with human related diseases, all labourers should practice strict personal hygiene and should know that their health may affect the safety of the end user.

Important rules to adhere to are listed below:

- Wash hands under running water and with soap after each visit to the bathroom.
- Wear a hair net if you are involved in packing and further processing of the produce.
- Wear a mask over the mouth and nose if you have to handle fruits and vegetables.
- In the case of a severe cold or other disease, inform your employer or group leader so that the correct health procedures can be followed.
- If you have or if you suspect you have some other contagious disease, inform your employer or group leader so that the appropriate measures can be followed. This may involve you being used in aspects of the harvest process which will not place you in direct contact of the produce etc. Remember your problem might be curable, but if you do not inform or trust your employer or group leader with this information, you could suffer unnecessarily.
- When you have open wounds on your body, cover it with clothes or bandages. Replace bandages at least once a day and make sure you are treating the condition correctly. If the wound is on the hands, use gloves when the produce has to be handled. Also inform your employer or group leader of this situation, to ensure the right health procedures are followed.
- In case of an accident during the harvesting or post harvesting of the crop, call the person(s) responsible for first aid to the scene of the accident, and then inform the group leader and employer immediately.
- If a fellow worker is not adhering to the rules set out by the employer, report it to the group leader or employer promptly. Rules are usually set to protect all role players in the harvesting process, this includes you, and a person not following these rules are not only endangering him/herself, but you also. Therefore, love yourself and report any potential problems promptly.
- Safety and the use of loaders and fork lifts

  The safe and efficient operation of any equipment, including loaders, is the sole and direct responsibility of the operator. Experience and skill in this type of task can only be gained through training and supervision.

  It is important that a safety inspection program is followed. Such inspections allow for the detection of unsafe equipment and unsafe conditions early. In this way counter measures can be set up before someone is injured. However accidents do occur and should someone be injured the injury must be reported immediately so that the injury can be properly treated and produce is not contaminated. This also allows for an injury record to be evident over time. Such records will highlight potential problem areas that can be eliminated from the harvesting procedures.

General safety inspections should be conducted daily before the equipment is being used. So checkpoints for equipment such as tractors and forklifts are:

- Check water, fuel, fan belts, etc.
- Once engine is running, check hand and air brakes, this ensures that brakes will hold while loading.

- Inspect hydraulic lines for kinks, cracks, and general wear and tear.
- Inspect the frame for hairline cracks, or fatigue in the metal. Cracks in the metal may cause the equipment to break either completely or parts coming off.
- Pickers must be at a safe distance from the equipment when loading or unloading bins.
- Operators must be aware of the location of the harvesters at all times. Drivers and operators must be on the look out for harvesters on ladders while driving through an orchard.
- Operators must look out for overhead power lines during loading and unloading of fruit.
- Always discuss unsafe acts seen during the day with the supervisors so that preventative measures can be developed and implemented.

Threshing

Threshing is the operation of separating the grains from the plants. These operations may be carried out in the field or on the threshing floor, by hand or with the help of animals or machines. Whatever the system used, it is very important that threshing be done with care. Otherwise, these operations can cause breakage of the grains or protective husks, thus reducing the quality of product and subsequent losses from the action of insects and moulds. Care should also be taken while transporting the harvested crops from the field to threshing floor to avoid any loss.

Hand Threshing

Hand threshing is generally done in case of pigeon pea. One of the simplest systems for threshing of pigeon pea crop is to strike the sheaves of crop spread over threshing floor with a flail or a stick. While this method can be adopted for other pulse crops like chickpea, mungbean, urdbean, fieldpea, rajmash and lentil but a more convenient method is threshing with animals or vehicles. The threshing-floors on which the sheaves are spread must have a hard, clean surface. By using method of hand-threshing, a worker can obtain 15-40 kg of product per hour.

Threshing with Animals or Tractors

Subject to the availability of draught animals, large quantity of crops can be threshed by treading the animals over about 30 cm thick layer of sheaves. This operation, which is also called "treading out", can equally well be accomplished with vehicles. Grain is obtained by running the tractor twice over sheaves of harvested and dried crops that are spread in layers on a circular threshing floor 15-18 m in diameter. The sheaves must be turned over between the two passages of the tractor.

Threshing with Hand Driven Machines

Normally a hand-operated machine like Olped thresher, which is basically used for threshing of paddy can also be used for threshing of cut stocks of pigeonpea. By means of the handle or pedal, a big drum fitted with metal rings or teeth is made to rotate. The cut-stocks of pigeonpea is threshed by hand-holding the sheaves and pressing the upper portion of dried plants against the rotating drum. The speed of the threshing drum must be kept at about 300 revolutions perminute (rpm). The hand-held sheaves must all be of the same length with the panicles all laid in the same direction, and the grains must be very ripe and dry. The machine must be continuously and regularly fed, but without introducing excessive quantities of product. Use of these threshing machines may require two or three workers. Depending on the type of machine, the skill of the workers and organization of the work, yields can be estimated at a maximum of 100 kg/h.

Threshing with Motorized threshers

The use of motorized threshers or threshers operated by tractor power has become very common for threshing of pulse crops. The threshers available in the market are basically designed for threshing of cereal crops and no specific thresher is available for threshing of pulse crops. However, by the simple replacement of a few accessories and the appropriate changes in settings, these machines can be used for threshing of pulse crops as well. Few popular designs of cereal threshers like Amar thresher, CIAE Multicrop thresher and Sonalika threshers have been successfully used for threshing of pulse crops with some degree of variations in the grain recovery. Table shows the principal technical features of these machines and the yield obtained in case of different crops. Equipped with a rotating threshing drum (with beaters or teeth) and a stationary counter-thresher, these machines often have devices to shake out the straw and to clean and bag the grain. Whether self-propelled or tractor-drawn, these threshers are often mounted on rubber wheels for easy movement to the field. The use of motorized threshers may require two or three workers. Yields depend on the type of machine, the nature and maturity of the grain, the skill of the workers and organization of the work.

Threshing Capacity of Different Threshers

Threshers	Technical details	Suitable for	Capacity (kg/h)
Sonalika	25 hp tractor, Peg type, Single blower	Chickpea, Lentil	300-350
Amar	7 hp motor, Peg type, Double blower	Pigeonpea, Chickpea, Urdbean	100-350
CIAE	7 hp motor, Peg type, Double blower	Chickpea, Lentil	300-450

Wind Winnowing

Winnowing is a farming method developed by ancient people for separating grain from chaff. It is also used to remove weevils or other pests from stored grain. Threshing, which is the loosening of grain or seeds from the husks and straw, comes before winnowing.

In its simplest form it involves throwing the mixture into the air so that the wind blows away the lighter chaff. The heavier grains fall back down for recovery. This method is called "wind-grading". Other methods include using a winnowing fan (a shaped basket shaken to raise the chaff). Yet another is using a tool (a winnowing fork or shovel) on a pile of harvested grain.

Winnowing can also describe the natural removal of fine material from a coarser sediment by wind or flowing water. This is similar to separating wheat from chaff.

Sorting/Food Grading

Sorting and grading are terms which are frequently used interchangeably in the food processing industry, but strictly speaking they are distinct operations. Sorting is a separation based on a single measurable property of raw material units, while grading is "the assessment of the overall quality of a food using a number of attributes". Grading of fresh produce may also be defined as 'sorting according to quality', as sorting usually upgrades the product. Virtually all food products undergo some kind of sorting operation. There are a number of benefits, including the need for sorted units in weight-filling operations and the aesthetic and marketing advantages in providing units of uniform size or colour. In addition, it is much easier to control processes such as sterilisation, dehydration or freezing in sorted food units; and they are also better suited to mechanised operations such as size reduction, pitting or peeling.

Criteria and Methods of Sorting

- Sorting is carried out on the basis of individual physical properties.
- Weight is usually the most precise method of sorting, as it is not dependent on the geometry of the products.
- Eggs, fruit or vegetables may be separated into weight categories using spring-loaded, strain gauge or electronic weighing devices incorporated into conveying systems.
- Using a series of tipping or compressed air blowing mechanisms set to trigger at progressively lesser weights, the heavier items are removed first, followed by the next weight category and so on.
- These systems are computer controlled and can additionally provide data on quantities and size distributions from different growers.
- An alternative system is to use the 'catapult' principle where units are thrown into
- Different collecting chutes, depending on their weight, by spring-loaded catapult arms.
- A disadvantage of weight sorting is the relatively long time required per unit; and other methods are more appropriate with smaller items such as legumes or cereals, or if faster throughput is required.

Size Sorting

- Size sorting is less precise than weight sorting, but is considerably cheaper.
- The size and shape of food units are difficult to define precisely. Size categories could involve a number of physical parameters, including diameter, length or projected area.
- Diameter of spheroidal units such as tomatoes or citrus fruits is conventionally
- considered to be orthogonal to the fruit stem, while length is coaxial. Therefore rotating the units on a conveyor can make size sorting more precise.
- The main categories of screens are fixed aperture and variable aperture designs.

- Flatbed and rotary screens are the main geometries of the fixed bed screen and a number of screens may be used in series or in parallel to sort units into several size categories simultaneously.
- The problem with fixed screens is usually contacting the feed material with the screen, which may become blocked or overloaded. Fixed screens are often used with smaller particulate foods such as nuts or peas. Variable aperture screens have either a continuous diverging or stepwise diverging apertures. These are much more gentle and are commonly used with larger, more delicate items such as fruit.

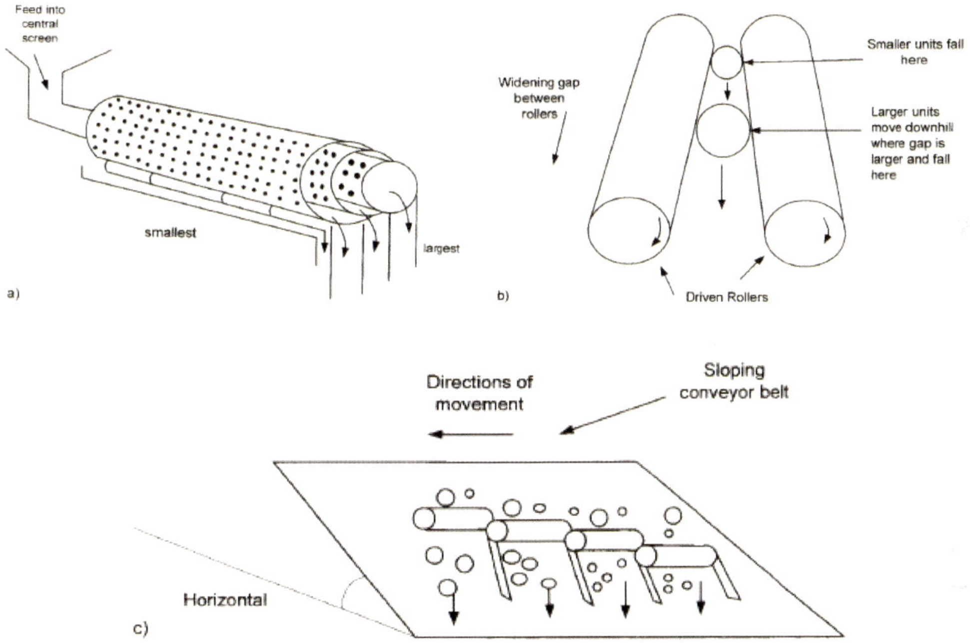

Figure: Geometries of size sorting equipment ;(a) concentric drum screen. (b) roller size sorter. (c) belt and roller sorter.

Shape Sorting and Density

- Shape sorting is useful in cases where the food units are contaminated with particles of s milar size and weight. This is particularly applicable to grain which may contain other seeds. The principle is that discs or cylinders with accurately shaped indentations will pick up seeds of the correct shape when rotated through the stock, while other shapes will remain in the feed.
- Density can be a marker of suitability for certain processes. The density of peas correlates well with tenderness and sweetness, while the solids content of potatoes, which determines suitability for manufacture of crisps and dried products, relates to density. Sorting on the basis of concentrations.

Photometric Properties

- Photometric properties may be used as a basis for sorting. In practice this usually means colour.

- Colour is often a measure of maturity, presence of defects or the degree of processing. Manual colour sorting is carried out widely on conveyor belts or sorting tables, but is expensive.
- The process can be automated using highly accurate photocells which compare reflectance of food units to preset standards and can eject defective or wrongly coloured, e.g. blackened, units, usually by a blast of compressed air. This system is used for small particulate foods such as navy beans or maize kernels for canning, or nuts, rice and small fruit.

Colour Sorting

- Extremely high throughputs have been reported. By using more than one photocell positioned at different angles, blemishes on large units such as potatoes can be detected.
- Colour sorting can also be used to separate materials which are to be processed separately, such as red and green tomatoes. It is feasible to use transmittance as a basis for sorting although, as most foods are completely opaque, very few opportunities are available. The principle has been used for sorting cherries with and without stones and for the internal examination, or 'candling', of eggs.

Grading

- Grading is classification on the basis of quality (incorporating commercial value, end use and official standards, and hence requires that some judgement on the acceptability of the food is made, based on simultaneous assessment of several properties, followed by separation into quality categories. Appropriate inspection belts or conveyors are designed to present the whole surface to the operator.
- Trained manual operators are frequently used to judge the quality, and may use comparison to charted standards, or even plastic models. For example, a fruit grader could simultaneously judge shape, colour, evenness of colour and degree of russeting in apples. Egg candling involves inspection of eggs spun in front of a light so that many factors, including shell cracks, diseases, blood spots or fertilisation, can be detected.
- Machine grading is only feasible where quality of a food is linked to a single physical property and hence a sorting operation leads to different grades of material. Size of peas, for example, is related to tenderness and sweetness, therefore size sorting results in different quality grades.
- Grading of foods is also the determination of the quality of a batch. This can be done by human graders who assess the quality of random samples of foods such as cheese or butter, or meat inspectors who examine the quality of individual carcasses for a number of criteria.
- Alternatively, batches of some foods may be graded on the basis of laboratory analysis.

Insect Damaged Product

Insect pests inflict their damage on stored products mainly by direct feeding. Some species feed on the endosperm causing loss of weight and quality, while other species feed on the germ, resulting in poor seed germination and less viability. Thus, due to damage done by insects, grains lose value for marketing, consumption or planting. Most storage pests are able to increase in numbers drastically within a relatively short time. At an early stage of development, population growth takes the "exponential" form, where the number of insects at a given time can be expressed by this equation: $N_t = N_o e^{rt}$, where N_t is the number of insects at time t, N_o is the original number of insects and r is the rate of intrinsic increase of the population. However, this pattern of growth would eventually reach an upper plateau due to depletion of food and intraspecific competition. In that case, the number of insects at any given time can be expressed by the equation:

$$N_t = K/(1+((K- N_o)/N_o)\, e^{-rt})$$

where, K is the maximum number of insects that the environment can support, which is also known as the "carrying capacity" of the environment. In grain stores, K is not constant and will decrease as the food is consumed. Eventually, the insect population will start to decline due to the decrease in food availability and competition among individuals, which is when a proportion of insects will have to migrate and search for other food sources.

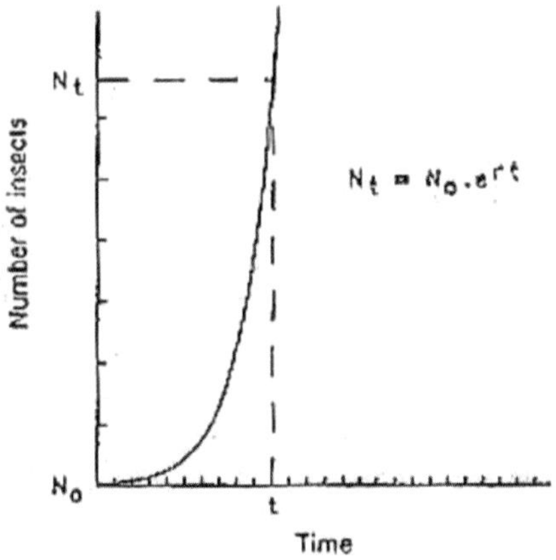

Figure: The theoretical exponential i ncrease of an insect population

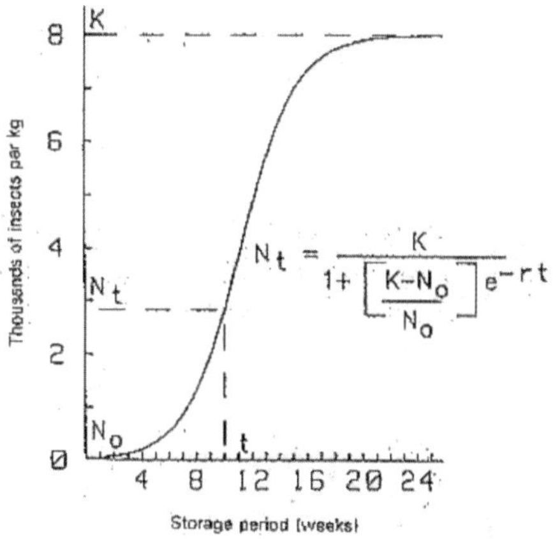

Figure: The logistic growth of an insect population in a restricted environment

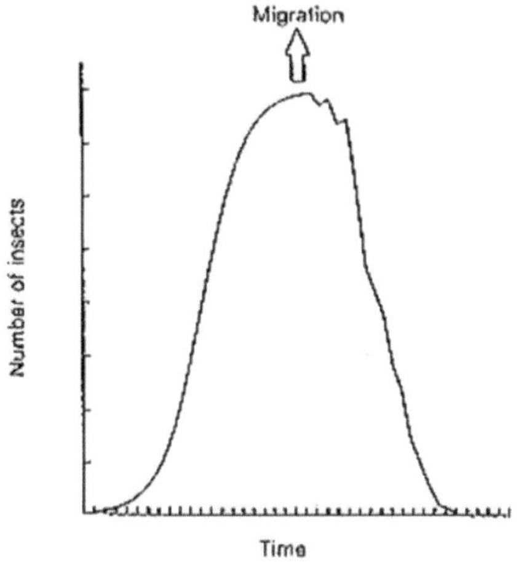

Figure: Insect population growth in a restricted environment without food replenishment

In addition to direct consumption of the product, insect pests contaminate their feeding media through excretion, moulting, dead bodies and their own existence in the product, which is not commercially desirable. Damage done by insect pests encourages infection with bacterial and fungal diseases through transmission of their spores. The presence of insects also raises the product temperature, due to their feeding activity, resulting in "hot spots". These spots in turn lead to concentrating of humidity within the product, thus stimulating seed deterioration and further fungal activity. There are a wealth of studies examining the effect of insect pests on grain contents. In Brazil, for example, showed that the presence of Sitophilus zeamais and Sitotroga cerealella in maize grains led to a reduction in germination with increasing developmental stage of the insects, from 13 percentage at the egg stage for Sitophilus zeamais and 10.9 percentage for Sitotroga cerealella, to 93 percentage and 85 percentage at the adult stage for S. zeamais and S. cerealella respectively) found that infestation of wheat, maize and sorghum grains with single or mixed populations of Trogoderma granarium and Rhyzopertha dominica resulted in substantial reductions in the contents of total lipids, phospholipids, galactolipids, and polar and nonpolar lipids recorded a significant decrease in essential amino acids in the same crops due to mixed infestation with the same two pests, with maximum reduction found in methionine, isoleucine and lysine. Similarly, recorded a substantial reduction in starch in parboiled cassava chips due to infestation with Sitophilus oryzae and Rhyzopertha dominica as compared to the uninfested chips. In Nigeria recorded high level of moisture, combined with a decrease in germination ability of maize due to infestation by Sitophilus zeamais, while showed that infestation of bambarra groundnuts with Callosobruchus subinnotatus reduced seed viability and increased free fatty acids and peroxides, which are indices used in measuring biochemical deterioration.

Major Insect Pests of Stored Foods

Two major groups of insects harbour the mostly economically important post-harvest insect pests: Coleoptera (beetles) and Lepidoptera (moths and butterflies). Several Coleopteran and Lepidopteran species attack crops both in the field and in store. Crop damage by Lepidoptera is only done by the larvae. Several lepidopteran larvae entangle the feeding media through silky secretion which turns products into entwined lumps. In the case of Coleoptera, both larvae and adults often feed on the crop and the two stages are responsible for the damage. Post-harvest insect pests may be primary, i.e. able to attack intact grains such as the genus Sitophilus, while others are secondary pests, attacking already damaged grains or grain products such as the genus Tribolium. The following is a list of the most common post- harvest and storage pests, their biology, distribution and common host plants.

Coleoptera

The order Coleoptera is the largest order of insects and contains the most common and important stored product pests. Adults have their forewings modified as hard elytra. Beetles inhabit a wide variety of habitats and can be found almost everywhere. Those associated with stored products exhibit different behavioural types; some are primary and secondary pests feeding directly on the product, others are general scavengers, fungus feeders, wood borers or predators of other insects. arvae lack the presence of prolegs (abdominal legs) and only possess true legs on the three thoracic segments. Larvae of a few species may also lack true legs, e.g. Sitophilus spp.

Curculionidae (Snout Beetles)

This is a large group of beetles that contains some of the most serious crop and stored grain pests. Members of this family are characterised by the form of the snout (rostrum) which is elongated in most species. This family contains the most destructive stored grains pests in the world.

The Rice Weevil: Sitophilus oryzae (L.) (=Calandra oryzae L.)

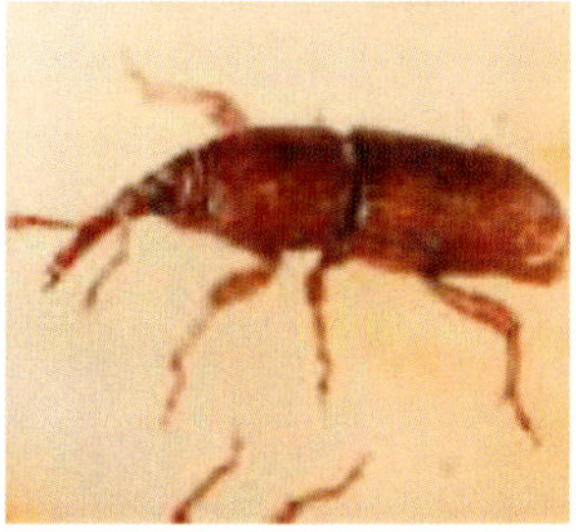

Figure: Rice weevil

The Maize Weevil: Sitophilus zeamais Motsch. (=Calandra zeamais Motsch.)

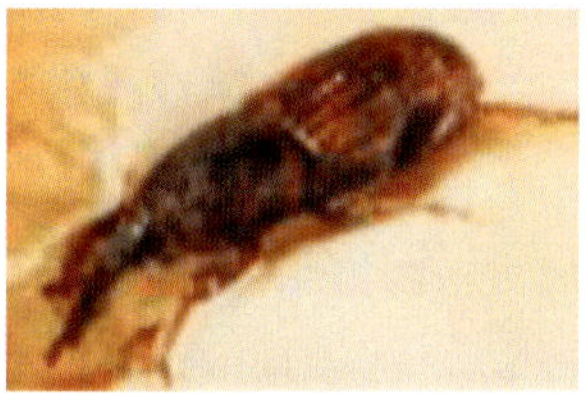

Figure: Maize Weevil

The Granary Weevil: Sitophilus granarius (L.)

The first two species are major primary pests that have a virtually cosmopolitan distribution throughout the warmer parts of the world. The rice weevil mainly attacks rice and wheat in stores, while S. zeamais is a serious primary pest of stored maize. However, both species are able to develop on all cereals, dried cassava and other processed food products.

The two species are morphologically identical. In Europe, the two species are replaced by the granary weevil, S. granarius, which is wingless and can be distinguished by the sculpturing on the prothorax and elytera.

The life cycle and damage caused by both S. oryzae and S. zeamais are similar. However, S. zeamais is a little larger (5 mm in length) and a very active flier. Infestation usually starts in the field and later continues in the store. Both species are capable of inhibiting reserved breeding grounds near the threshing floors that are normally full of plant residues, where the population builds up in before moving to granaries. Adult females chew grains creating a small hole in which they lay eggs and then seal the hole with a secretion. The optimum temperature for oviposition is around 25oC and at grain moisture contents of over 10 percentage. Larvae tunnel in grains and are responsible for most of the damage. pupation takes place inside the grain and adults chew their way out through the outer layer of the grain. Adults live for 5-6 months depending on the temperature and humidity of grains.

S. oryzae adult females can lay more than 500 eggs during their lifetime. The optimal temperature for development is 30°C with maximum oviposition taking place at 18 percentage humidity. The rice weevil can live without food for 6-32 days depending on temperature. This species is highly affected by changes in temperature; all stages die in about a week at 0° C. On the other hand, S. zeamais tolerates lower temperatures than S. oryzae and can live for 37 days at 0° C.

The granary weevil, S. granarius, lives for one full year at 20-25° C and a relative humidity of about 15 percentage. Biology of this species is similar to the other two species, but it is unable to fly, thus restricted to the store. This species prefers softer grains such as wheat, rye and barley, as food and habitat. In addition, S granarius has a high resistance to low temperatures; adults can stay alive for up to two months at -5° C. Insects can be controlled if exposed to 50° C for 35 minutes which will kill all stages. showed that S. oryzae is displacing S. granarius in Switzerland.

Tenebrionidae (Darkling Beetles)

This is a large and varied group of insects that contains more than 10,000 species of which about 100 are associated with stored products. Most of the tenebrionids are black or dark brown in colour and mainly phytophagous. Adults are characterised by the tarsi of the hind leg with only four segments. Infestation by these beetles results in an unappealing smell due to the secretion of benzoquinones from abdominal glands. The following tenebrionids are serious secondary pests of stored grains and flour.

The Red-Rust Flour Beetle: Tribolium castaneum Herbst.

Figure: Red-Rust Flour Beetle

The Confused Flour Beetle: Tribolium confusum

These two species are probably the most common secondary pests of all plant commodities in store throughout the world. Several other species of Tribolium are occasional minor pests and can be found in almost every store containing infested cereals or cereal products, specially in tropical and sub-tropical climates. Both species attack maize, wheat, flour and other foodstuffs, but T. confusum does not seem to be as common as T. castaneum in tropical climates. Members of genus Tribolium are known to produce toxic quinones which contaminate flour and flour products. Damage is done by both larvae and adults specially to broken or damaged grains.

T. castaneum adult females lay small, cylindrical, white eggs scattered in the product. At an optimum temperature of 32.5° C, females lay up to 11 eggs daily. Larvae are yellowish with a pale brown head, and they live inside grains until pupation. Adults are about 3-4 mm long and can live for a year or more. Females are highly fecund and able to lay a maximum of 1000 eggs during a lifetime, with 40° C and 22° C as upper and lower limits for development. This species is also highly tolerable to humidity as low as 11 percentage. Adults are highly adapted to feed on a very wide range of commodities and perfect colonizers of new habitats. In tropical conditions, this species is dominant to T. confusum.

The confused flour beetle, T. confusum, is often confused with T. castaneum but they can be separated using the last three segments of the antenna which are much larger than the rest in

T. castaneum and forming a club, while the last five segments in T. confusum gradually enlarge towards the tip. Just like T. castaneum, the confused flour beetle develops in crushed grain products and a constant inhabitant of flour mills specially in the temperate regions of the world. In contrast to T. castaneum, this species is not able to fly, but has a long life span that can reach three years under moderate climatic conditions .

The Yellow Mealworm Beetle: Tenebrio molitor

Tenebrio beetles are black or dark brown and they feed as larvae and adults on grain products. T. molitor is an important post-harvest pest and occurs spread all over the world. Adults are elongate, 16 mm long, and active fliers. Females can lay up to 600 eggs during its lifetime. Larvae firstly eat the germs of stored grains and can feed on a wide variety of plant products such as ground grains, flour, tobacco and foodstuffs. Larvae are very voracious and highly resistant to low temperature; they can remain alive for 80 days at -5° C.

Other tenebrionids are less common polyphagous pests around the globe such as T. destructor, T. madens and Palorus depressus.

Bostrichidae (Branch and Twig Borers)

Members of this family are elongate with the head bent down ventrally to the thorax. Adults are characterised by rasp-like hooks on the pronotum. Most of the species are borers in wood or roots. Wood boring activities of these beetles may weaken timbers or wooden walls of the stores. This family contains two serious stored grain pests:

The Lesser Grain Borer: Rhizopertha dominica (Fabricius)

The lesser grain borer (R. dominica) attacks a wide range of stored cereals. It can be found attacking cassava, flour and other cereal products and is also able to attack rough rice grains. The pest originated from South America, but is now found in all the warmer parts of the world. This species is a serious pest in Australia, from where it was carried to the USA and other parts of the world during World War I. Adults of this species are tiny dark beetles, 2-3 mm in length, and are very voracious with a long life span. Females may continue to lay eggs for four months and are able to lay up to 500 eggs at 34°C. They feed externally on grains and lay eggs on their surface. Larvae feed either externally or inside the grain and pupation takes place within the eaten grain. Larval development is relatively faster when fed on whole grains than on flour. Both adults and larvae eat the endosperm leaving powdered grains. This dust can accumulate on the walls of the warehouses and it is a sign of high infestation. Though are not common on pulses, adults are able to breed in grains that are too dry for fast development of Sitophilus. At 34°C, development is possible on grains with moisture contents as low as 9 percentage, and they can daily destroy grains equal to their body weight.

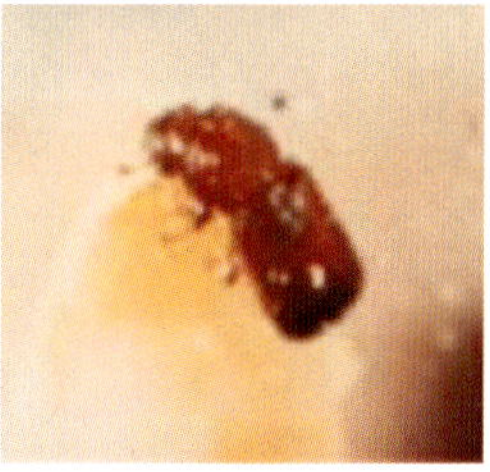

Figure: Lesser Grain Borer

The Larger Grain Borer: Prostephanus truncatus (Horn)

The larger grain borer is a primary pest, often attacking maize in the field towards the end of the season, then continuing in the store. P. truncatus is a serious pest of maize in Central America and many parts of Africa. It was first reported in East Africa in 1981 and in 1984 in West Africa. Since then, it has spread rapidly in the African continent where it has become a major pest of stored maize and dried cassava. In Togo, soon after the discovery of P. truncatus, mean losses of 30.2 percentage were reported on stored maize six months after storage. Stored dried cassava is also known to become heavily infested by P. truncatus, which may lead to cross infestation of maize. Hodges et al. reported 70 percentage loss in dried cassava roots after four months of storage due to this species.

Figure: Larger Grain Borer

Adults of P. truncatus bore in maize grains and produce large quantities of dust, in which their larvae seem to feed and pupate. This species proved to be highly tolerable to low moisture contents in grains. Field studies in Tanzania recorded heavy infestation in maize at a moisture content as low as 9 percentage. The introduction of this pest in Africa has influenced the economy of several countries, specially those depending on exporting of maize. Many countries now refuse to import maize from areas infested with the larger grain borer.

Bruchidae (Seed Beetles)

Most bruchids are short, stout-bodied beetles with a short forewing not reaching the tip of the abdomen. Adults are characterised by their compact hairy bodies and relatively long antennae. Larvae of most species feed inside seeds and some develop in stored dry grains or legumes. All bruchids are phytophagous with most species able to avoid feeding on seed covers that contain toxins. This family contains several important field and stored crop pests.

The Cowpea Weevil: Callosobruchus maculatus (Fabricius)

This is an important pest that mainly attacks beans of various species, and can alternatively attack other pulse crops. This species originated in Africa but is now found all over the tropics and sub-tropics. Adults are 2-4 mm, brownish with black markings. They have a short life span of about 12 days and do not feed. Two forms of this species have been identified; the active (flying) form and the flightless form. The flying form disperses and colonises cowpea fields. Adult females lay about 100 eggs glued to the seed surface or to pods. Larvae tunnel inside the seed where the entire development takes place. In the store, the normal form continues to reproduce until the end of the storage season. The flying form appears again in response to disperses to new locations. This species causes major problems in Nigeria and Niger, where most of Africa's cowpeas are produced. Other species such as C. rhodesianus and C. subinnotatus may also be important in some parts of Eastern and Central Africa.

The American Bean Weevil: Acanthoscelides obtectus Say (Bruchus obtectus Say)

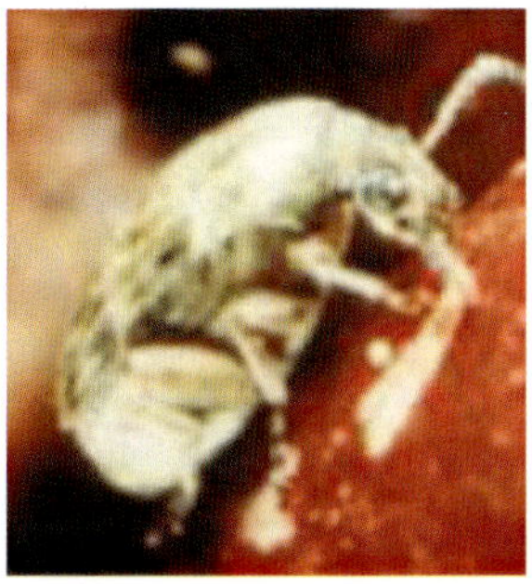

Figure: American Bean Weevil

This species is widely distributed in Africa, Central and South America, New Zealand, USA and Southern Europe. A. obtectus exhibits high tolerance to varied degrees of temperature, thus, it is found in cool highland areas as well as the warmer parts of the tropics. It mainly attacks beans of various types and other pulse crops. Adults are grey and oblong in shape, with the body covered by yellowish green hairs. Females are almost twice as large as males. Infestation starts in the field when females lay eggs on the mature beans in plant pods. Larvae are tiny with strong mandibles and feed inside the seeds where life cycle is completed. Adults exit the seed through round holes about 2 mm in diameter.

The Groundnut Borer (Seed Beetle): Caryedon serratus (Olivier) = (C. gonagra (F)).

This species is common in West Africa and parts of South Eastern Asia where it probably originated. Adults are 4-7 mm in length with distinct serrate antennae. C. serratus attacks mainly groundnuts and other legumes, pods and seeds of Acacia tress and tamarind. Adult females glue their eggs on groundnut seeds soon after harvest. Larvae bore inside seeds making a large hole in the cotyledon. Pupation may take place inside or outside the kernel in paper-like cocoon attached to the pod. C. serratus is a serious pest of stored groundnuts in West Africa.

Several other bruchids are known as post-harvest pests in different geographical areas of the world:

Species	Remarks
Callosobruchus chinensis (Linnaeus)	Originated in tropical Asia, but is currently distributed all over the tropics and sub-tropics. It attacks chickpeas, cowpeas and green grams. Life cycle and damage is very similar to C. maculatus.
Callosobruchus subinnotatus (Pic)	Formerly described as a strain of C. maculatus. It is found in West Africa where it attacks "Bambarra groundnuts"
Callosobruchus theobromae (Linnaeus)	A pest of pigeon pea in India and was recorded in a groundnut field in Nigeria
Bruchidius atrolineatus (Pic)	Mainly a field pest of cowpeas but eggs and larvae are taken to storage after harvest recorded 90% pods infestation from Niger in West Africa

Bruchus atomarius L.	Distributed in Europe and parts of Asia. Attacks beans, peas and lentils
Bruchus lentis Frol	A monophagous species that occurs in some warmer parts of the world. This species infests lentil seeds in stores.
Bruchus pisorum L.	Reported from Europe, Canada, South East Asia and former USSR. A monophagous species that attacks ripe plant pod and can only develop on peas.
Bruchus rufipes Herbst	Distributed in central and southern Europe, Asia and south Africa. Attacks vetch seeds in which they develop.
Bruchus dentipes Baudi	This species occurs in bean cultivating area. Infests seeds of broad beans and other species of the genus Vicia.

Cucujidae (Flat Bark Beetles)

Members of this family are small flattened beetles, mostly found under the bark of trees or in tunnels made by other beetles. This family contains one common pest of stored grains.

The Red Rust Grain Beetle: Cryptolestes ferrugineus (Stephens)

Adults of this species are oblong flattened small beetles (1.5-2 mm long), with the head and prothorax relatively big and conspicuous. C. ferrugineus is a widespread secondary pest of stored grains, specially in the humid tropics. The genus Cryptolestes was reported to be of economic importance towards the end of the maize storage season in Togo. However, it might not be as serious as other pests in stores, often following an infestation by other insects. It usually attacks the germs of broken or cracked grains thus reducing germination. Other species such as C. pusillus (Schonherr) and C. pusilloides (Steel and Howe) are common in humid areas of the tropics.

Silvanidae

This family was formerly included in Cucujidae. It includes two important species:

- The Saw-toothed Grain Beetle: Oryzaephilus surinamensis (L), recognized by the toothed lateral margins of the pronotum.
- The Merchant Grain Beetle: Oryzaephilus mercator (Fauvel), which is found in association with O. surinamensis.

Both species are virtually cosmopolitan and they infest a wide variety of stored grains, processed foodstuff and other food products. They are mainly secondary on stored products following more destructive primary pests. However, O. surinamensis prefers cereal products while O. mercator is more frequent on oil-seed products and more temperature sensitive.

They enter damaged grains and feed specially on the germ. Optimum conditions for development are between 30- 35°C and 70-90 percentage relative humidity.

Adults are 3 mm flattened narrow winged beetles but they rarely fly. Females lay their eggs loosely within the stored products. Larvae are free living and start by feeding on the embryo and the endosperm. They require 60-90 percentage humidity for optimal development, and neither species cannot develop or breed at temperatures less than 19° C. All stages die in ten minutes if exposed to 55° C.

Dermestidae (Skin Beetles)

Members of this family are ovoid in shape with hairy or sometimes scaly bodies. Larvae are very hairy. When stores are infested, these setae may be seriously hazardous if inhaled by workers. This family contains a number of very destructive and economically important species. One of the most serious stored product pests that belongs to this family is the Khapra Beetle: Trogoderma granarium Everts. Apparently the only phytophagous species in the genus Trogoderma. A native of India, the Khapra beetle is now found in most parts of the world specially hot and dry areas. Adults are oval, red brown insects with a dark thorax.

Adult females may lay up to 120 eggs within the stored products. Larvae are considered primary pests as they attack undamaged grains and seeds and bore into stored pulses. They are highly mobile, and in the absence of food they enter a diapause that might last for more than two years, in which they can be highly resistant to the application of pesticides or fumigation. Adults are 3-4 mm long, dark wingless beetles that do not feed. Populations of this pest build up rapidly, specially in the hot humid tropics. This species was apparently eradicated in the United States and the former Soviet Union. It also seems to be absent from East and southern Africa.

Anobiidae

Anobiids are cylindrical pubescent beetles, 1-9 mm in length. The head is usually concealed from above by the hoodlike pronotum. Most anobiids live in dry vegetable materials or bore in wood, while others are fungus feeders. About 1000 species of Anobiidae are known, most of which are found in the tropics. The following are two widespread storage pests belonging to this family.

The Cigarette Beetle: Lasioderma serricorne (Fabricius) is a common pest of stored cereals, cocoa beans, tobacco, ground nut, peas, beans, flours and other foodstuffs. Originally from South America, it is now found in most of the warmer parts of the world. This species is notorious for attacking a wide range of intact cereal grains, pulse seeds and food stuffs.

Adults can breed anywhere at optimum temperatures of around 28-32° C and a relative humidity of 75 percentage. Newly hatched larvae are very active and responsible for most of the damage. Adults are small brown beetles and the only damage they cause is due to their emergence holes. This pest can be controlled if exposed to temperatures below 18° C. At 55° C, all stages die in two hours.

The Drug Store Beetle: Stegobium paniceum (Linnaeus) Another widespread pest that infests several cereals, but less common than L. serricorne in the tropics.

Adults are 3.5-4 mm in length, brown hairy beetles, and they do not feed. Females lay about 75 eggs and optimum conditions are 300C with 60-90 percentage relative humidity. Larvae are active feeders and they can be indiscriminate in their food choice, biscuits, macaroni, dry fruits and other products. This species is commonly found in the temperate areas of the world.

Trogossitidae (Bark Gnawing Beetles)

Trogosittids are brownish beetles. The Caddle (Tenebroides mauritanicus (L.)) is a common pest in granaries. Observed for the first time in Mauritius, it is now considered a cosmopolitan pest associated with a wide variety of commodities. T. mauritanicus attacks mainly cereals, oilseeds and their products. Both adults and larvae are highly tolerant to very cold conditions. Though larvae are known to predate upon other insects, both adults and larvae feed directly on stored food and larvae are able to tunnel in wooden walls of the store to create a pupation chamber.

Lepidoptera

Lepidoptera is the second most important order of insects pests of stored products. Adults are active flyers with two pairs of scaly wings. Mouthparts of the adults are modified to suck plant nectar or other fluids and are not able to chew, while those of the larvae possess well- developed mandibles. Larvae are distinguished from beetle larvae by their pseudopods (false legs) on some of the abdominal segments. Lepidoptera larvae occur frequently in a wide range of habitats and are known for their silk-spinning activities that result in the additional loss of quality of stored products. Some species attack the product in both the field and store. Several moths are pests of the ripening crop and their larvae can be found in recently harvested stored grains. They either continue their attack for a short time in the store or form an entry point for further attack by true storage pests.

Pyralidae

Pyralidae is a large family, of which only a few species are stored product pests. Most pyralids are small and delicate moths. Members of this family exhibit a great deal of variation in appearance and habits. Larvae of all species possess glands which secrete silk with which they interlink food products as they move. This family is divided into a number of subfamilies, with the subfamily Phycitinae containing some of the most important stored grain pests. The best-known species in this subfamily are the following:

The Tropical Warehouse Moth: Ephestia cautella (Walker) = Cadra cautella Hb.

A very serious cosmopolitan stored product pest infesting a wide variety of hosts such as maize, wheat, and other grains in stores. It can also feed on dried fruits, beans, nuts, bananas and groundnuts.

Adult females lay up to 300 small round sticky eggs within the substrate and through holes in bags. Optimum conditions for larval development are 32-33°C and 70 percentage relative humidity. Larvae feed on the seed germ and are fairly mobile within the produce. A considerable amount of damage results from webbing in the grain and on the surface of bags forming large lumps, therefore food is no longer fit for consumption once infested. Pupation takes place in crevices or between bags. Adult moths spread the infestation in the warehouse through egg laying. This pest is cosmopolitan in tropical and sub-tropical parts of the world.

The Warehouse Moth: Ephestia elutella (Hub.)

This species is a polyphagous pest that feeds on a vast variety of stored products such as dried cocoa beans, dried grains, pulses, nuts, tobacco, coconut and dried fruits. Infestation is mainly post-harvest.

The whole life cycle takes about 30 days at 30°C and 70 percentage relative humidity. Most of the damage is due to contamination of food with exuviae, dead bodies and frass. Silk produced by larvae may be extensive. The warehouse moth is a world wide pest, but more abundant in the sub-tropics and temperate areas of the world. This pest shows high levels of resistance to several groups of pesticides.

The Mediterranean Flour Moth: Ephestia kuehniella = Anagasta kuehniella (Zeller).

Adults are similar to E. cautella but the body is relatively longer. A major pest of flour mills, its main habitats are flour and grout mills, corn milling plants, bakeries and any other place used for processing grains or preparing flour products. E. kuehniella occurs in most of the temperate and sub-tropical parts of the world, where average temperatures are around 20° C- 25° C. Complete development requires about 74 days at 25° C and 75 percentage relative humidity. Larvae entwine all the material on which they feed resulting in solid lumps of food particles, faeces and larval exuviae.

Indian Meal Moth: Plodia interpunctella (Hübner)

This insect feeds mainly on meals and flours but can attack raisins, nuts and some pulses and whole cereals. The Indian meal moth is distributed all over the tropics and sub-tropics and in some parts of the temperate regions, specially in heated buildings. In the hot tropics, it is more abundant in cooler highland areas. Most of the damage occurs due to larval feeding on the germinal part of the grains. Damage also occurs through the contamination of foodstuff with dead larvae, frass and silk webbing.

Larvae feed in tubes they weave from silk secretions. Adult females stick about 200-400 eggs to the substrate or to the storage walls. Larvae develop and feed within the substrate and are sensitive to changes in temperature. The number of generations may be only two per year in Europe, but increases in the tropics to eight generations. Complete development takes about 27 days at 30° C and 70 percentage relative humidity. Development ceases below 15° C. All stages die at 55° C in five hours.

Gelechiidae

Gelechiidae is a large family of lepidoptera. All moths are small in size and several species are important plant pests. This family contains two serious post-harvest pests:

The Angoumois Grain Moth: Sitotroga cerealella (Olivier)

Figure: Angoumois Grain Moth

This species is a serious primary pest that mainly attacks maize, wheat and sorghum, both in the field and in stores. A recent survey in southern Ethiopia revealed that this pest alone was responsible for 11.2 to 13.5 percentage weight loss in stored maize. Infestation with S. cerealella starts in the field as females lay their eggs, singly or in groups, on grains. Larvae start feeding inside the grains, while still in the milk stage, and spend their entire life inside one grain. Thus, infestation is difficult to detect at this stage. Adults leave a conspicuous emergence hole at one end of the kernel. Infested grains are characterised by this circular window created by the larvae. Stored grains may be completely destroyed. Adults are active fliers, thus, they are able to infest neighbouring granaries, which is known as "cross- infestation". This pest is distributed throughout the warmer parts of the world (Africa, South and Latin America and southern Asia and Australia).

The Potato Tuberworm: Phthorimaea operculella (Zeller) = (Gnorimoschema operculella (Zeller))

This species is a cosmopolitan pest of potatoes, tomatoes and eggplants. It attacks plants mainly in the field, but continues to feed on tubers in storage. Larvae mine in the leaves and stems and later bore into the tubers. Damage can be seen on leaves as silver spots due to the tunnelling larvae, or as tunnels in the plant stem.

Each female lays about 150-200 eggs and larvae tunnel through leaves and stem down to the tuber where pupation may take place. In the store, eggs are laid individually on the tubers near the eyes or on sprouts. P. operculella is an important pest in traditional potato stores in North Africa. High infestations of up to 50 percentage of tubers can take place in Yemen due to this pest.

Acaridae

The Flour Mite: Acarus siro

Mites are widely distributed tiny arthropods. They can live and develop on various plant in the field or indoors. Mites can be found in granaries, feed mixing plants, threshing floors, stacks of hay and straw, dead organic matter, soil or plant residues. Several species are predacious on other mites or insects. Mites are easily transmitted by virtue of their tiny size which allows them to be carried with dust, winds, insects, birds or rodents. About 30 mite species are known to be associated with stored products. Family Acaridae contains some damaging species, in which Acarus siro is probably the most important and commonly encountered mite in granaries. This mite is about 0.7 mm in length with an oval body. A. siro is a widely distributed polyphagous species that can be found on almost all products of plant or animal origin. It requires relatively high humidity (70 percentage), with humidities below 11 percentage being lethal to the mite. Temperatures below -15° C for 24 hours kill all stages. At 60° C, all stages die in 5 minutes.

Attacked grains lose nutrients and the ability to germinate due to feeding on the germ. Crushed bodies of Acarus cause coloration in flour that reduces the products value. Under normal conditions, this mite develops according to the following pattern: egg, larva, nymph I, nymph II, and adult. Some strains of A. siro may produce hypopus under favourable conditions. Hypopus is a diapause form that can be carried by rodents or insects to other storing places. However, this species does not seem to occur in most of the tropical lowlands, though it might sometimes infest grains in cooler upland areas.

Fungal Contamination and Production of Mycotoxins

Another important cause of grain deterioration is infection by fungal diseases. Just like infestation by insect pests, fungal infection mainly starts in the field and is later carried to the store. High relative humidity is a crucial factor for encouraging fungal infestation. Factors influencing the degree of humidity in the store can be a high moisture content in the product if it has not sufficiently dried after harvest, infestation with insect pests that results in hot spots and increased humidity, or improper storing technique that allows for contact with rain water or humidity condensation. Fungal infestation results in reduction of grain quality, change in colour, taste, smell, reduction in nutritional value, increase in free fatty acids (FFA) and reduction of germination ability.

Fungal Diseases

Fungal diseases may be highly hazardous as certain species of fungus produce mycotoxins, which are poisonous substances produced by moulds during their growth and development. Mycotoxins are highly stable compounds that cannot be destroyed through food processing, and the only way to avoid them is to prevent the fungal growth. The first recorded case of poisoning due to food contamination with fungal infestation was in the early 1930s, when 5000 farm horses died in Illinois, USA, due to a disease that was called the "mouldy corn disease". It occurred among farm animals that fed on maize left in the field after harvest. Later in the mid 1930s, a plant pathologist in Minnesota, USA, isolated Fusarium sp. from maize infected by ear rot, and the extract gave similar disease symptoms on swine. A few years later in the former USSR, hundreds of people were affected by what was later described as "alimentary toxic aleukia". People had eaten millet from plants overwintered in the field and gathered later in the spring. The grains were infected by different species of fungi, some of which produced potent toxins. In 1960, about 100,000 turkeys died in England of an unknown disease. Later a fungus identified as Aspergillus flavus was isolated from a suspected groundnut meal that was imported from Brazil. Extracts from this fungus confirmed the presence of a toxic substance that was given the name "aflatoxin". This material has been extensively studied and proved to be highly toxic to man and farm animals. It is a liver toxin which can induce cancer in susceptible animals. Fungal growth can be very rapid as well as the production of aflatoxin, specially in tropical and sub-tropical countries, where environmental conditions are highly conducive.

The fungus is widely distributed all over the world and has been found on all foodstuff and their products. Several strains of A. flavus produce aflatoxin and can contaminate grains, pulses, cassava, oilseeds and other foodstuff. Factors during cereal storage can favour the development of the fungus and the production of aflatoxin. A moisture content that is slightly above 9 percentage in groundnuts or around 16 percentage in cereals is enough to support the development of the fungus.

Aflatoxin has been found in sausages in Germany and other meat products. In the Philippines, aflatoxin was found in the majority of the samples of peanut butter in stores. Moreover, aflatoxin consumed by dairy cattle, though altered in their body, still remains toxic and shows up in the milk.

A. flavus can grow and produce aflatoxin in many kinds of plants and plant products. However, major agricultural crops in which aflatoxins can create a serious problem are groundnut, maize and cottonseeds, specially where crops are grown in warm and humid conditions. On the other hand, not all strains of A. flavus produce aflatoxins, some can even be used in the preparation of foods for human consumption. Several other Aspergillus species and other fungi in different genera are associated with stored products, some of which may produce other important mycotoxins.

The following is a list of the most common stored product fungus species:

Aspergillus Candidus

This fungus is common in stored grains and their products where moisture content is at least 15-16 percentage. It is known to cause preliminary heating of stored grains. Its presence is an indication that a stored lot is contaminated with spoiled grains.

Aspergillus Clavatus

This fungus is commonly found in soil and decaying plant materials. It requires a moisture contents of 23-25 percentage in cereal seeds and can grow at lower relative humidities on groundnut meal or copra.

Aspergillus Fumigatus

This fungus occurs in decaying plant materials and requires relatively high temperature to develop (40° C). It was reported to result in a high level of abortion in cattle feeding on contaminated food. A. fumigatus may also infect human lungs. However, this species requires a high relative humidity of 95-100 percentage to grow.

Aspergillus Parasiticus

An aflatoxin producing fungus which attacks maize, groundnuts and oilseeds.

Aspergillus Restrictus

This species is known to have a "restricted" growth. It is able to kill and discolour wheat germ at a narrow range of relative humidity of 13.8-14.3 percentage. A. restrictus is usually associated with rice weevils, but even when the insect pest is eliminated, the fungus will continue to grow. It is also associated with some grain infesting mites.

Alternaria Alternata

An important mycotoxin producing fungus that attacks rice, sorghum and soybeans.

Fusarium Graminearum

This species produces deoxynivalenol, which is a serious and acute human toxin. It also produces zearalenone. Both toxins are produced on maize, wheat and barley.

Fusarium Moniliforme

This species commonly invades stems of maize plants and it is known to produce the mycotoxin, fumonisin. In high moisture conditions, F. moniliforme may be involved in rotting of the kernel. However, it requires a 22 percentage moisture contents to grow, thus, it does not cause serious problems in stores.

Fusarium Roseum

This species causes scab of wheat, barley and oats. Symptoms are the discoloration of seeds. It also causes "ear rot" in maize and may continue developing on maize left on the plants after harvest.

Fusarium tricinctum

A mycotoxin producing fungus. Heavy infestations are common when maize is stored on the cobs in cribs.

Helminthosporium spp.

Fungi belonging to this genus may cause seed infections in different cereals such as maize and rice.

Scopulariopsis spp.

A predominant fungus associated with black and white pepper, soybean flour and powdered milk.

Penicillium verrucosum

This fungus infects barley and wheat. It produces ochratoxin, a mycotoxin that may lead to kidney damage in farm animals. Other certain species of Penicillium produce citrinin, an important mycotoxin that may lead to kidney damage in humans and farm animals.

Fungal Infestation

Improper handling of crops during post-harvest processes can cause fungal infestation. Any damage to stored products increases their susceptibility to fungal contamination. More importantly, insects activity can have a profound effect on the spread of fungal diseases through transmitting the spores and increasing the surface area susceptible to fungal infection, which eventually increases the production of mycotoxins.

Dunkel indicated that some storage insect species are disseminators of storage fungi while others are exterminators; some storage fungi attract storage insects and promote their population increases while others repel and secrete toxins harmful to insects. Therefore, knowledge of basic biological relationships between insects and fungi in the stored grain ecosystem is crucial for their management. Several studies demonstrate the importance of insect pests as promoters or facilitators of fungal infection. In Nigeria, for example, showed that the yam beetle, Heteroligus meles, which was the largest cause of damage to tubers, facilitated the spread of different Fusarium species and other less abundant fungi. None of the fungi was able to infect undamaged yams in the laboratory. In India, sampled wheat grains for Sitophilus oryzae and found that A. flavus was the most frequent species in their alimentary canals, followed by A. candidus, A. sojae, A. fumigatus, Penicillium rugulosum and Cladosporium cladosporioides. This indicates the possibility that S. oryzae transmits fungus spores from infested to healthy grains. In the USA, Beti showed that maize kernels infested with A. flavus- contaminated Sitophilus zeamais weevils had higher levels of aflatoxin than A. flavus- inoculated maize without weevils. The presence of S. zeamais resulted in increased kernel moisture content which was positively correlated with aflatoxin contents. In addition, aflatoxin levels in infested maize increased with increasing numbers of A. flavus- contaminated S. zeamais, as S. zeamais carried spores both internally and externally on their exoskeleton.

In some crops, efforts to remove broken and discoloured seeds can effectively reduce the production of mycotoxins. However, this may not be practical for many products, especially when the fungal growth is internal and difficult to detect. In Thailand, an in-store drying system to control aflatoxin contamination in maize was developed. High moisture maize is dried to 18 or 19 percentage moisture content within 2 days and continuously dried to 14 percentage within 14 days. An airflow rate of 3.6-4.6 m^3/min per m^3 of maize is required to decrease moisture content from 19 to 12 percentage. In an experiment in India, cinnamon oil treatment of maize, in combination with sodium chloride, synergistically inhibited fungal infection, growth and aflatoxin production.

Contaminated Product

Contamination can occur at any point along the chain—during production, processing, distribution, or preparation.

Production

Production means growing the plants we harvest or raising the animals we use for food. Most food comes from domesticated animals and plants, and their production occurs on farms or ranches. Some foods are caught or harvested from the wild, such as some fish, mushrooms, and game.

Production means growing the plants we harvest or raising the animals we use for food.

Examples of Contamination in Production

- If a hen's reproductive organs are infected, the yolk of an egg can be contaminated in the hen before it is even laid.
- If the fields are sprayed with contaminated water for irrigation, fruits and vegetables can be contaminated before harvest.
- Fish in some tropical reefs may acquire a toxin from the smaller sea creatures they eat.

Processing

Processing means changing plants or animals into what we recognize and buy as food. Processing involves different steps for different kinds of foods. For produce, processing can be as simple as washing and sorting, or it can involve trimming, slicing, or shredding. Milk is usually processed by pasteurizing it; sometimes it is made into cheese. Nuts may be roasted, chopped, or ground (such as with peanut butter). For animals, the first step of processing is slaughter. Meat and poultry may then be cut into pieces or ground. They may also be smoked, cooked, or frozen and may be combined with other ingredients to make a sausage or entrée, such as a potpie.

Processing means changing plants or animals into what we recognize and buy as food.

Examples of Contamination in Processing

- If contaminated water or ice is used to wash, pack, or chill fruits or vegetables, the contamination can spread to those items.
- During the slaughter process, germs on an animal's hide that came from the intestines can get into the final meat product.
- If germs contaminate surfaces used for food processing, such as a processing line or storage bins, germs can spread to foods that touch those surfaces.

Distribution

Distribution means getting food from the farm or processing plant to the consumer or a food service facility like a restaurant, cafeteria, or hospital kitchen. This step might involve transporting foods just once, such as trucking produce from a farm to the local farmers' market. Or it might involve many stages. For instance, frozen hamburger patties might be trucked from a meat processing plant to a large supplier, stored for a few days in the supplier's warehouse, trucked again to a local distribution facility for a restaurant chain, and finally delivered to an individual restaurant.

Distribution means getting food from the farm or processing plant to the consumer or a food service facility like a restaurant, cafeteria, or hospital kitchen.

Examples of Contamination in Distribution

- If refrigerated food is left on a loading dock for long time in warm weather, it could reach temperatures that allow bacteria to grow.
- Fresh produce can be contaminated if it is loaded into a truck that was not cleaned after transporting animals or animal products.

Preparation

Preparation means getting the food ready to eat. This step may occur in the kitchen of a restaurant, home, or institution. It may involve following a complex recipe with many ingredients, simply heating and serving a food on a plate, or just opening a package and eating the food.

Preparation means getting the food ready to eat. This step may occur in the kitchen of a restaurant, home, or institution.

Examples of Contamination in Preparation

- If a food worker stays on the job while sick and does not wash his or her hands carefully after using the toilet, the food worker can spread germs by touching food.
- If a cook uses a cutting board or knife to cut raw chicken and then uses the same knife or cutting board without washing it to slice tomatoes for a salad, the tomatoes can be contaminated by germs from the chicken.
- Contamination can occur in a refrigerator if meat juices get on items that will be eaten raw.

Mishandling at Multiple Points

Sometimes, by the time a food causes illness, it has been mishandled in several ways along the food production chain. Once contamination occurs, further mishandling, such as undercooking the food or leaving it out on the counter at an unsafe temperature, can make a foodborne illness more likely. Many germs grow quickly in food held at room temperature; a tiny number can grow to a large number in just a few hours. Reheating or boiling food after it has been left at room temperature for a long time does not always make it safe because some germs produce toxins that are not destroyed by heat.

Culling

Culling is elimination or weeding out of undesirable animals from the herd, for reasons of uneconomic,, poor production, or very poor reproductive ability, with sterility problems and breeding, irregularities, very poor conditions, stunted growth, suffering from incurable illness, or disease animals found to be positive for serious infections diseases like Tuberculosis, Johnes disease, Brucellosis, lost one or more quarters and teats of the under due to chronic mastitis resulting in marked reduction in milk production.

Undesirable breed characters present in young animal. When the herd is a pure bred herd leading to disqualifications family lines, exhibiting heritable characters like supernumerary teats, loose horns in cows of certain breeds.

Disable animals due to injury or loss of organ, extreme lameness leading to unmentionable conditions, un healed fractured animals etc. come under the animal proposed or culling.

The culled animals carry lower values and a separate list is made for such called animals and it is known as culling list.

When the culling cows for poor production, the entire lactation at yield is considered and preferably first two lactations are observed and if the lactation yield is less than what is expected from the breed or herd, The animal is included in the culling list. Very old animals are culled, as their maintenance will be uneconomical.

Male animals or other animals surplus in the farm or not useful in the farm and they are culled. Calves born with congenital defects like congenital fibrosis of the eye, total blindness or some other defects are included in the culling. Calves born much below the normal birth weight are included in the culling. Yearlings animals male or females, stunted much below their normal body weight, potbellied conditions bad confirmation are culled. Dairy Animal Management Valuation and culling is done on the farms every year at least once in year. In some farms culling is done twice a year however doing it once a year is must.

Tips for Culling

- Do not move non-ambulatory animals to market under any circumstances.
- Make the decision to treat, to cull, or to euthanize animals promptly. Sick and injured animals should be segregated from the herd.
- Delay transport of an animal that appears to be exhausted or dehydrated until the animal is rested, fed, and dehydrated.
- Milk all cows that are still lactating just prior to transporting to a packing or processing facility.
- Use a transportation company that is knowledgeable about your animal care expectations and provides for the safety and comfort of the animals during transport.
- Do not transport animals to a packing or processing facility until all proper treatment withdrawal times have been followed.
- Do not transport animals with a poor body condition, generally a Body Condition Score of less than 2 (1-5 scale).
- Do not transport animals that require mechanical assistance to rise and are reluctant or unable to walk, except for veterinary treatment. When using any handling device, abuse must not be tolerated.

- Do not transport animals with bone fractures of the limbs or injuries to the spine. Animals with a recent fracture unrelated to mobility should be culled and transported directly to a packing or processing facility.
- Do not transport animals with conditions that will not pass pre-slaughter inspection at a packing or processing facility. If unsure, consult with your veterinarian before transporting an animal to a packing or processing facility.

Curing of Roots

Curing is a postharvest treatment of crops to reduce water loss and decay during storage. In root and tuber crops, curing refers to the process of wound healing with the development and suberization of new epidermal tissue called wound periderm. In bulb crops, curing refers to the process of drying of the neck tissues and of the outer leaves to form dry scales. Crops can be cured in the field or in facilities designed for the process.

uring root and tuber crops such as sweet potatoes, potatoes, cassava and yarns is an important practice if these crops are to be stored for any length of time. Curing is accomplished by holding the produce at high temperature and high relative humidity for several days while harvesting wounds heal and a new, protective layer of cells form. While curing can be initially costly, the long extension of storage life makes the practice economically worthwhile.

The best conditions for curing vary among crops as shown in the following table:

Commodity	Temperature		Relative Humidity	Days
	C	F	(%)	
Potato	15-20	59-68	90-95	5-10
Sweetpotato	30-32	86-90	85-90	4-7
Yarns	32-40	90-104	90-100	1-4
Cassava	30-40	86-104	90-95	2-5

Curing, when used for onions, garlic and flowering bulbs refers to the practice directly following harvest, of allowing the external layers of skin and neck tissue to dry out prior to handling and storage. If local weather conditions permit, these crops can be undercut in the field, windrowed and left there to dry for five to ten days. The dried tops of the plants can be arranged to cover and shade the bulbs during the curing process, protecting the produce from excess heat and sunburn. If forced heated air is used for curing onions and other bulbs, one day or less at 35 to 45° C (95 to 113° F) and 60 to 75% relative humidity is recommended. The dried layers of 'skin then protect the produce from further water loss during storage.

Field Curing

Yams and other tropical root and tuber crops can be cured outdoors if piled in a partially shaded area. Cut grasses or straw can be used as insulating materials and the pile should be covered with canvas, burlap or woven grass mats. Curing requires high temperature and high relative humidity, and this covering will trap self-generated heat and moisture. The stack should be left for about four days.

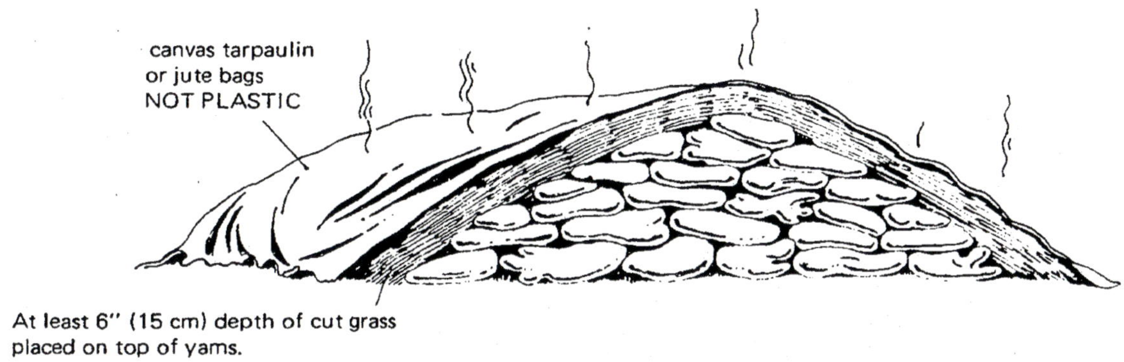

Onions and garlic can be cured in the field in regions where harvest coincides with the dry season. The crops can be cured either in windrows or after packing into large fiber or net sacks. The produce can be left in the field for five days, then checked daily until the outer skin and neck tissues are properly dried. Curing may take up to ten days, depending on weather conditions.

Curing can be assisted by the use of ventilated sheds in regions where solar radiation and/or relative humidity is high or natural air movement is low. Produce in sacks can be stacked in the shade on canvas tarpaulins under one or more ceiling fans.

Curing Assisted by Shade and Ventilation

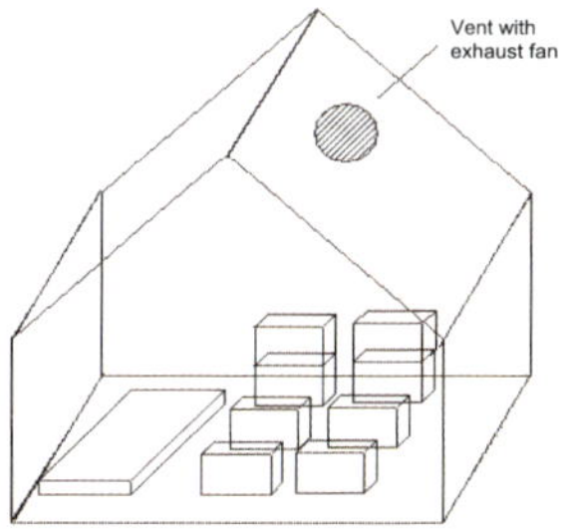

Curing with Heated Air

The most uniform distribution of heat is obtained when heat is introduced near the floor level of a curing structure. Heaters can be placed on the floor near the bins of produce, or heat can be ducted in from outside the curing room. A high relative humidity can be obtained by wetting the floor or by using an evaporative cooler in the room without introducing outside air.

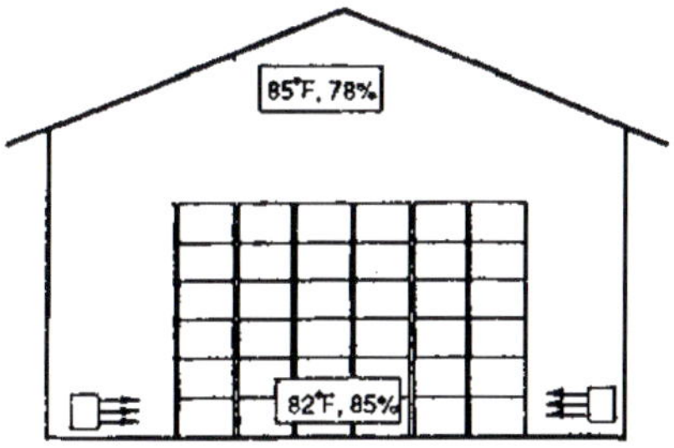

If heaters are located near the ceiling, then ceiling fans can be used to help redistribute the heat down into the room of produce. Bulk bins must be stacked to allow a gap of 10 to 15 cm (4 to 6 inches) between rows for adequate air circulation.

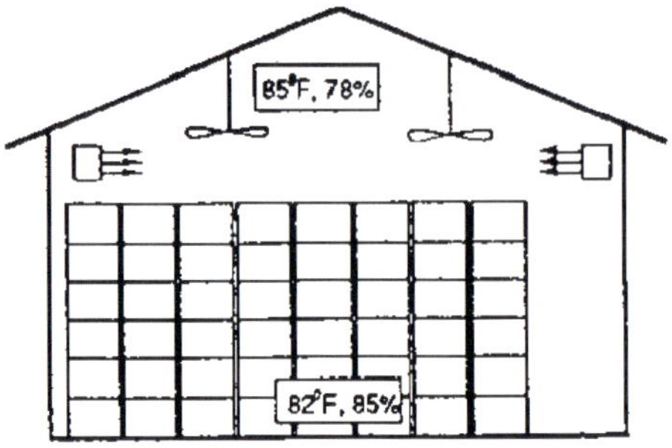

Bulk Systems for Curing Onions

Curing using a bulk system requires a fan, a heating unit and a slatted floor. The illustrations below shows how air can be brought in, heated and distributed through a load of bulk onions in the curing room An exhaust opening near the ceiling recirculates heated air.

When using heated air it is easy to overdry the bulbs, leading to loss of external scales and exposure of the fleshy scales underneath. Curing onions should be checked regularly to avoid overdrying.

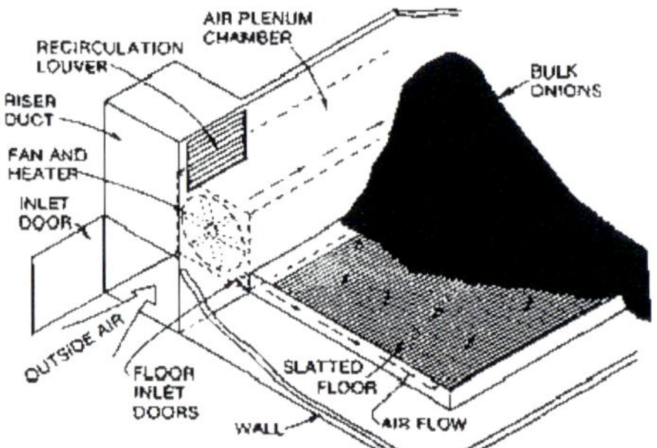

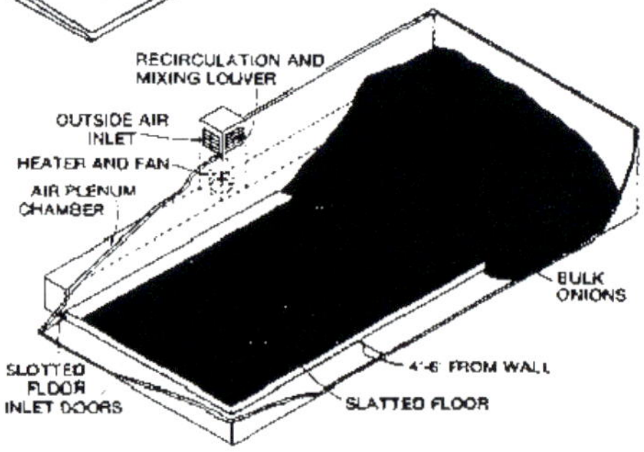

Emergency Curing

If conditions such as rain or flooded fields do not permit field curing and curing facilities are not available, a temporary tent can be used for curing onions. In the example illustrated below, the tent is constructed from large tarps. Heated air is forced into a hollow area (known as a plenum) at the center of the bins of produce. Several fans are used to circulate the warm air through the onions while they are curing.

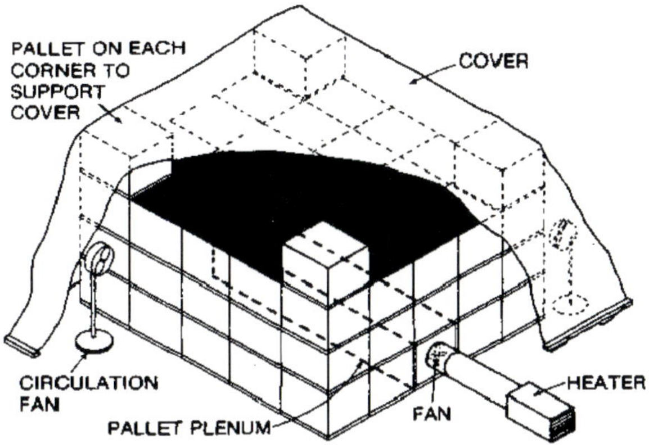

Opta comnimincias qui occab il im et eos earuntur

References

- Storage-crops, vegetables, growers-library: johnnyseeds.com, Retrieved 28 May 2018
- Harvesting, food-and-cooking, food-and-drink, sports-and-everyday-life: encyclopedia.com, Retrieved 18
- Harvesting-processes, step-by-step-production: knowledgebank.irri.org, Retrieved 25 June 2018
- Threshing, post-harvest-management-of-pulses, technologies-for-agri-horti-crops, post-harvest-technologies, agriculture: vikaspedia.in, Retrieved 05 July 2018
- Animhus-culling-dairy, animal-husbandry: agritech.tnau.ac.in, Retrieved 27 April 2018

2

Postharvest Losses

Post-harvest losses occur between the stages of harvest and human consumption. Losses of horticultural produce and grains are significant in the post-harvest sector. This chapter has been carefully written to provide an easy understanding of the varied aspects of postharvest losses in grains, fruits and vegetables. It includes vital topics such as vegetable pathogens, senescence in fruits and fruit pathogens.

Postharvest Losses in Fruits and Vegetables

Fruits and vegetables are highly important as they provide a balanced and healthy diet to people. Though adequate quantities of fruits and vegetables are being produced all over the world, yet on account of losses in the field as well as in storage, they become inadequate. Generally, about 30 % fruits and vegetables are rendered unfit for consumption due to spoilage after harvesting.

Post harvest loss of fruits and vegetables has been defined as "that weight of wholesome edible product (exclusive of moisture content) that is normally consumed by human and that has been separated from the medium and sites of its immediate growth and production by deliberate human action with the intention of using it for human feeding but which for any reasons fails to be consumed by human." Not only quantity and quality but even the appearance of fruits and vegetables are affected and their market value is reduced. Most 'skin deep' injuries such as 'fly speck' in apples do not affect the edible part. Some infections are not harmful as they occur on inedible parts and can be trimmed off before consumption ego Alternaria spots on the outer leaves of cabbage.

Fresh fruits and vegetables are perishable and highly prone to these losses because they are composed of living tissues. These tissues must be kept alive and healthy throughout the process of marketing. These are composed of thousands of living cells which require care and maintenance.

Therefore, the reduction of post-harvest loss of fruit and vegetables is a complementary means for increasing production. It may not be necessary to considerably step up the production of fruits and vegetables with the growing demand if the post-harvest loss is reduced to a great extent. The cost of preventing losses after harvest in general is less than preventing a similar additional amount of fruit and vegetable crop of the same quality. Attention to the concept of post-harvest food loss reduction, as a significant means to increase food availability, was drawn by the World Food Conference held in Rome in 1974.

- Factors Affecting Post-Harvest Loss

  Several factors influence the post-harvest losses in fruits and vegetables. These include losses due to physical, physiological, mechanical and hygienic conditions. Fruits and vegetables are characterized by high metabolic activities and known to possess short storage life. Due to these factors, significant losses occur between harvesting to consumption.

Certain other factors like insect and mite injury, diseases due to non-infectious pathogens and pathological rots are also responsible for the post-harvest loss of fruits and vegetables. Among the causes, the pathological rots are the most serious followed by mechanical injury. Pathological rots together with mechanical injury cause a lot of damage to the perishables.

Environmental factors such as temperature, relative humidity and oxygen balance especially in storage are also greatly responsible. But the loss of fruits and vegetables are closely interrelated with the loss due to physiological and biochemical deterioration. Environmental conditions are also responsible for rendering fruits and vegetables for pathological attacks. Among these, temperature and humidity are more important.

Therefore, post-harvest loss of fruits and vegetables can be studied as primary and secondary causes.

- Primary causes

  Primary factors of post-harvest loss in fruits and vegetables are those for which certain mechanical, physiological, pathological or environmental factors are directly responsible. They are described as follows:

  - Mechanical injury

    Mechanical loss to the fruits are caused by careless handling during harvesting, packing, transportation, storage etc. Some insects and birds are also responsible for the mechanical injury in fruits and vegetables. Mechanical injuries like bruising and cracking of fruits and vegetables render them more prone to attack by organisms and significantly increase the rate of water loss and gaseous exchange. Many a tines the mechanical injury received by fruits and vegetables due to the pressure thrust during transportation, though sometimes invisible, leads to the rupturing of inner tissues and cells. Such produce is degraded faster during the natural ageing process.

    It is very likely that fruits and vegetables receive maximum mechanical injury during harvesting if proper means of picking and harvesting are not adopted. A large quantity of tuber and root crops like potato, sweet potato, etc. become unmarketable due to faulty methods of digging them out. Among fruits, certain fruits like aonla, apples, etc. have soft outer skin, which are highly prone to mechanical damage. Fruits thus damaged are also prone to pathological attacks like rotting etc. Gradually the quality of fruits also deteriorates and the produce fetch lesser prices in the market.

    Processing operations such as spillage, abrasion, excessive polishing, peeling and trimming add to the loss of commodity. Puncturing of the containers and defective seals also leads to mechanical injury.

  - Microbial action

    Like any other food, fruits and vegetables are also prone to microbial spoilage caused by fungi, bacteria, yeast and moulds. A significant portion of losses of fruits and vegetables during post-harvest period is attributed to diseases caused by fungi and bacteria. The succulent nature of fruits and vegetables makes them easily invaded by these organisms. Besides attacking fresh fruits and vegetables, these organisms also cause damage to canned and processed products. Many serious post-harvest diseases occur rapidly and cause extensive break down of the commodity, sometimes spoiling the entire package. It is estimated that 36 % of the vegetable decay is caused by soft rot bacteria. Similarly fruit rot in aonla and other soft fruits caused by fungi is also very destructive. As far as vegetables are concerned, naturally the source of infection is from the field, water used for cleaning the surface, contact with equipment and storage environment. The most common pathogens causing rots in vegetables and fruits are fungi such as Alternaria, Botrytis, Diplodia, Monilinia, Phomopsis, Rhizopus, Pencillium, Fusarium, etc. Among bacteria Ervinia, Pseudomonas, etc. cause extensive damage.

High temperature and relative humidity favour the development of post-harvest decay organisms. More acidic tissue is generally attacked by fungi, while fruits and vegetables having (pH) above 4.5 are more commonly attacked by bacteria, ego bacterial soft rot of potato caused by Ceratocystis, fimbriata, water soft rot of carrot by Sclerotinia sclerotiorum etc.

- Environmental factors

Among environmental factors, temperature, humidity, composition and proportion of gases in controlled atmospheric storage etc. play an important role.

High temperature and relative humidity favour the growth of micro-organisms which cause extensive damage to the produce. High temperature also increases the rate of respiration in fruits and vegetables which subsequently leads to the breakdown of the inner tissues. Decaying of fruits and vegetables is increased by high temperature coupled with high relative humidity. The microbial attack to different crops becomes very slow at low temperature especially when it is below 5° C.

Chilling injury by low but not freezing temperature is commonly observed with tropical and subtropical fruits and vegetables. The symptoms of chilling injury may not be evident while the fruits and vegetables are held at chilling temperature but become noticeable only after they are transferred to room temperature (22° C).

Role of relative humidity (RH) in the post-harvest i environment is often as important as temperature. Effects of temperature and relative humidity are often comparable and interrelated because the capacity of air to hold moisture varies with the temperature. Aeration in the storage containers or in stores has its bearing on RH and hence indirectly on disease development. Effects of high RH on decay is closely related with the effects of temperature and, for many fruits and vegetables, relative humidity near saturation results in lower decay losses only if the temperature is near O° C. Relative humidity below 90 % does not permit micro-organisms to grow on the surface of fruits and vegetables.

Each fruit and vegetable has specific heat requirement during processing. Excessive or insufficient heat during processing, improper cold storage temperature and un- desirable gaseous composition of controlled atmosphere of storage lead to physical damage due to tissue break down. Hollow heart disease in potato is a good example of faulty oxygen balance in storage or during transportation. A large chunk of potatoes is destroyed by this disorder every year.

- Secondary causes

  Inadequate harvesting, transportation, storage and marketing facilities and legislation lead to the conditions favourable for secondary causes of loss. Inadequate harvesting facilities and rough handling during harvesting results in bruising and increase possibilities of contact of the produce with the soil which leads to contamination with organisms. A prolonged period taken for harvesting and grading in field leaves the produce with field heat for longer time which, subsequently causes faster senescence. Besides, use of improper machinery and equipments in mechanical harvesting cause serious losses. For example, mechanical harvesting of tomato has been reported to result in more cracking (33 %) than hand picking (10 %).

  Harvesting of fruits and vegetables during rain or immediately after rains creates conditions favourable for decay organisms. Harvesting during hotter part of the day results in faster senescence, shrivelling and wilting of fruits and vegetables as compared to those harvested in the early morning or later afternoon.

  Inadequate storage facilities at the producing or marketing centres, leaves the produce to the natural causes of losses i.e. decay by organisms, respiration, transpiration and other biochemical reactions. Proper storage thus creates conditions unfavorable to these factors. Transportation and distribution of fruits and vegetables are the important areas of post-harvest loss. Physical and mechanical injuries occur during transportation and distribution. Longer shipment and distribution period eventually cause, heavy losses. The earlier the harvested produce is consumed, the minimum is the loss, as the period of senescence I and organism invasion, multiplication and damage is shortened. This can only be achieved by a well set up marketing system.

  The presence or absence of legal standards can affect the eventual retention or rejection of a vegetable or fruit for human consumption. Legal standards vary from country to country and are influenced by the economic status of the population and pressure of the population on the fruit or vegetable consumption. Strict legal standards will give better prices to farmers and superior quality to the consumers.

Control of Post-harvest Loss

The magnitude of post-harvest loss in fruits and vegetables can be minimized by proper cultural operations, harvesting, transportation, storage and pre and post- harvest treatments. These are briefly discussed here,

- Cultural operations

  Cultural operations which ensure the normal development of fruits and vegetables and avoid infection of decay organisms will help to prolong their storage. In case of vegetables, especially root crops such as carrot and radish, preparation of the soil to a fine tilth of porous nature is necessary to avoid root forking. Regular irrigation during the development of bulbs, tubers and fruit vegetables is of vital importance. Interrupted supply of water causes cracking of carrot, radish, tomato and splitting of outer scales of onions. Sudden and heavy irrigation at late maturity after a long dry spell, results in cracking of water- melon and tomatoes. Such specimens have short storage life and poor appeal. In onion and garlic, to ensure better keeping quality, irrigation should be stopped before" three weeks of harvesting. Irrigation of orchards during fruit bearing period is important to minimize fruit drop as well as fruit cracking especially in pomegranate.

Heavy application of nitrogenous fertilizers cause faster tissue deterioration in fruits and vegetables, while essential supply of potassic fertilizers improves the keeping quality of fruits and vegetables. Deficiency of molybdenum and boron in certain fruits and vegetables e.g. aonla, almond and cabbage leads to certain physiological disorders like greening of almond and heart rot of cabbage. Manganese deficiency in pea causes marsh spots. Such specimens have very poor storage life and may have a marked deterioration of quality too. Therefore, it is advisable to fulfill the optimum need of these micro-nutrients at appropriate time.

Effective weed control in orchards and vegetable fields increases the essential supply of water, nutrients, sunlight and consequently the normal development of fruits and vegetables. Proper training and pruning of fruit trees and removal of dead or infected branches enable the, trees to receive proper sunlight which is essential for the colour and size development of the fruits.

In case of vegetables, mulching should be practiced I as it conserves water, nutrients and keeps control over weeds and helps to obtain quality commodities. Besides, mulching obstructs the reflection of soil heat and the produce has a longer keeping quality. Operations such as staking i.e. (giving a support) in tomato, pea and other vine crops are useful in avoiding the contact of vegetables with soil and thereby reduces decay subsequent to harvest. Special operations like blanching (the curd is protected against sun-burning and yellowing by covering the curd with leaves etc.) of cauliflower gives quality curd that store longer in storage.

Menace of birds in the orchards can be controlled by using certain mechanical devices and careful supervision. Shooting, making drum sound, installing bird-scarers over trees, etc. are certain measures to avoid damage of fruits by birds.

- Harvesting and field handling

Consumers always prefer fresh, properly matured, insect and disease free fruits and vegetables of attractive appearance. Vegetables should be harvested as and when they attain the maximum size and yet are tender. Over maturity in root crops such as carrot, radish etc. causes sponginess and pithiness and therefore, their harvesting should not be delayed. The delayed harvesting of onion and garlic reduces their storage quality. Almost all the crop are badly affected due to delayed harvesting. Ripe cucurbitous vegetables are not preferred by the consumers. In case of fruits, ripening in trees is seldom prefer- red. Most of the fruits are harvested at proper maturity stage but quite earlier than the ripening stage. They are ripened artificially later on as per the market demands.

Harvesting of fruits and vegetables should be done at cooler parts of the day and the produce should be shifted to the packing shade as early as possible. Harvesting during hot period raises the field heat of the produce and consequently wilting and shriveling may occur. Harvesting of vegetables during or immediately after the rains should not be carried out as it creates conditions most favourable for the multiplication of micro-organisms.

In most of the developing countries, manual harvesting using knife and spade clippers for fruit picking is still in vogue. While harvesting vegetables, care should be taken to avoid mechanical injury. For large scale harvesting of tuber crops such as potato or the underground vegetables, mechanical diggers like potato digger can be safely used. In case of fruit harvesting, a number of harvesting aids have been developed to eliminate the use of ladders, for ego mobile platforms, which place the picker in the correct harvesting position. The large size of fruit trees poses a great problem in manual harvesting, resulting in more damage to the fruits. However, manual harvesting is easier and more efficient in the case of high density orchards where fruit trees planted are dwarf.

Buckets, sacks, baskets and boxes are used to collect the produce from the field to a central collection point. Many of these containers are poorly designed, inadequately maintained and unsuitable for job. The field container should be designed in such a fashion that they do not cause any injury to the field operations, a high standard of field hygiene should be maintained. Produce unfit for the market should be removed. Under no circumstances should the rejected produce be allowed to remain on the ground for any extended period of time as it becomes a source of infection. All the disease infected material should be completely destroyed.

- Washing and grading

Almost all the fruits and vegetables require special preparation before they are packed. Vegetables such as root and tuber crops are often washed to remove the soils adhering over them. Washing of fruits before packing is also very important, especially of those which have been treated with poisonous chemicals. Besides cleanliness, water used for washing improves the appearance of fruits and vegetables and prevents their wilting.

Many vegetables need to be trimmed before they are: ready for the market. Rotten, diseased, insect damaged and discolored leaves of vegetables such as cabbage, spinach, lettuce and several others are trimmed off. Part of the foliage of root crops, when bunched for market is removed by stripping off older leaves, and in some cases by cutting them back. Leafy greens such as green onion, spinach and fenugreek are tied in bundles. During washing, the organisms present on the fruits and vegetables are removed.

Grading is an important practice for successful marketing of fruits and vegetables. The produce is gene- rally sorted out into different grades and attractive forms to fetch a good price from them. Fruits and vegetables are graded according to shape, size and colour. Some fruits and yegetables are graded according to their maturity (okra, cucumber, ridge gourd etc.), ripeness (mango, tomato, etc.) and general appearance as well. In sorting the fruits and vegetables, all the characters that influence " their appearance and quality should be considered.

- Packaging

Packaging of fresh fruits and vegetables has a great significance in reducing the wastage. Packaging also provides protection from mechanical damage, undesirable physiological changes and pathological deterioration during storage, transportation and marketing. Through proper packaging, freshness, succulence and flavour of fruits and vegetables can be maintained for a longer period.

A wide variety of containers such as wooden boxes, bamboo baskets, jute bags, earthen pots and corrugated fibre board boxes are the important packaging materials used in most of the developing countries. Cheap packaging technique and materials such as polythene films, paper board boxes lined with polyethylene and other materials can effectively prolong the storage- life of fruits and vegetables. Vegetables can be protected from the influence of dry air by packing in various types of plastic films. Polyethylene. packaging provides a modified atmosphere and can consequently reduces decay, softening and loss of solids as observed in tomato, cauliflower, brinjal and okra.

All the packages must be provided with some amount of ventilation in order to prevent the physiological break- down of the produce. If there is impermeability to carbon dioxide (CO_2) and water vapour inside the package, there may be the development of off-flavour.

While packing for transportation, care should also be taken to avoid bruising and damage of fruits. For this purpose dry grasses or paddy straw can be successfully used between the layers of fruits.

These days, corrugated fibre boxes for the packaging of fruits are getting popular. They have many advantages:

- They are light in weight;
- They cause much less damage to fruits;
- They are easy to handle and print;
- They improve the image of the product;
- They reduce the freight cost; and
- They can be made from cheaper wood and other plant cellulose waste.

However, packaging is a highly innovative technology and newer techniques and material are being added every year, both for fresh vegetables and fruits and also for processed products. Packages can be flexible or rigid, each group having distinct advantages and limitations. An important application of flexible film is in wrapping the produce for low temperature. A mixture of gases can be inserted in film packs that delay post-harvest changes. Tetraking R, a D-shaped cylinder formed from two webs of polystryene and expanded polystyrene material has been recently introduced along with thermo formed pots for aseptic filling. Sterilization is carried out by either using liquid or vapour of Hydrogen peroxide or using ultra violet or gamma radiation.

- Pre-cooling

  It is desirable to remove field heat of the harvested vegetables, particularly when harvested during hot weather. This prevents the ripening and ageing of the produce. Prompt cooling conserves weight which gives an added advantage during the extended period of storage. In tomato, for eg. physiological loss in weight during storage can be reduced from 6 to 2.9 per cent. Vegetables such as artichokes and okra which deteriorate fast, need prompt cooling and their storage losses can be reduced by pre- cooling. Cooling is equally useful in case of fruits. It reduces their respiration rate especially, climatic fruits such as banana, papaya and mango, thereby preventing over-ripening. Sometimes, stages of ripening and the level of field heat also determines the need of pre-cooling. For eg., unless tomato fruits are kept above 26.7° C and ripening is to be delayed, there is no need for pre-cooling.

Pre-cooling can be accomplished by (1) placing the produce in refrigerated trucks with forced humidified air circulation, (2) placing ice in packages, (3) placing ice in water and passing through a spray of cool water, and (4) through vacuum cooling. Hydro-cooling, accomplished by flooding, spraying or immersing, is a rapid and affective method since water is an excellent material to transfer the heat from the produce to the cooling medium. A minute amount of an effective fungicide can also be added in hydrocooling water to control the fungal growth. However, the usefulness of hydrocooling is often questioned for large fruits or vegetables such as brinjal. Risk of water spotting is also evident.

Highly perishable leafy vegetables and salad crops are often cooled by reducing the atmospheric pressure in hermetically sealed chamber until the reduced vaporizing point of water produced at low pressure in the cooling chamber cool the produce. The technique is termed as vacuum cooling and its outstanding advantages are speed and uniformity in cooling of the commodity. Leafy vegetables and some fruits like grapes are especially difficult to cool with water or air, but they can be field packed and cooled rapidly and uniformly by vacuum cooling.

- Transportation

Transportation and distribution of the fruits and vegetables are the most important areas of post-harvest loss. In local market, the produce is brought either by bullock cart or tractor trollies. The long distance transportation is mainly by rails and trucks, which is very costly. The basic reason for preference to road transportation is because it takes short transit period.

Quick transport of fruits and vegetable in order to maintain the quality with minimum damage during transportation is very important for successful marketing of the produce. The losses in transportation occur due to physical and mechanical injury and uncontrolled conditions, mainly temperature and humidity. To ensure fewer losses, adequate long distance facilities especially by rails should be developed. The produce should be properly packed and stacked in well ventilated carriages. Refrigeration and relative humidity control in carriages is of vital significance. This can be achieved without using high cost mechanical refrigeration, simply, by adopting the technique of evaporation cooling with a slight modification in the existing road trucks and railway wagons. The surface of roads and their maintenance need improvement in order to have smooth transportation without severe jerks and vibration. The road transport can even become faster if all national highways are made one way and unnecessary delay caused by the closure of the level crossing is prevented. Hence, efficient transport system can go a long way, not only in reducing the post-harvest loss of horticultural produce but also in stabilizing the price fluctuations of the same commodity available in different markets of the country.

- Reducing moisture loss

Prevention of weight loss from fruits and vegetables is a major advantage of consumer packaging when moisture retentive films are used, particularly for leafy green vegetables and root crops. As a result, storage life of packaged produce may be lengthened over non-wrapped produce, particularly when marketing from refrigerated display cases. It has been reported that moisture proof cellulose film, reduced the weight loss of tomatoes to one fourth the amount lost by non-wrapped tomatoes. Lettuce packed in polythene or polystyrene, loses fewer weight and may have a longer storage life than non packaged lettuce.

Regulation of accumulating humidity is far more essential for packaged vegetables. Humidity attaining saturation is detrimental to onion, potatoes and sweet potatoes as ventilation with holes of one fourth to half inch sizes are a way of regulating package humidity to improve the keeping of onions and other produce.

- Marketing

  It is very important that horticultural produce reaches the market as soon as it is possible and at a time when the market needs it the most. A perfect and efficient marketing system is essential to avoid the losses of fruits and vegetables, and also to get a good return from the same. Marketing of perishable horticultural products presents more problems compared to other durable agricultural products. It is not efficiently managed in most of the developing countries. The interests of producers as well as consumers are poorly served, the grower gets less return and the consumer pays more than what is necessary.

  Sometimes, the market is in glut with a particular vegetable or fruit and a lot of wastage or loss is experienced. Also, prices are considerably reduced and farmers get dejected. Therefore, glutting should be avoided and the market should be so facilitated as to reduce the loss to the minimum.

- Low temperature storage

  Storage is one of the most important aspects of post- harvest handling of fruits and vegetables. The main objective of storage is to extend the storage life of fruits and vegetables and increase their period of availability. A substantial quantity of fruits go waste in our country due to lack of proper storage facilities. It should be known that fruits and vegetables are living entities and they carry out all the vital physiological activities even after harvest. The primary purpose of storage is to control the rate of respiration, transpiration, ripening and also any un-desirable bio-chemical changes and disease infection. The loss of many fruits and vegetables can be prevented to a great extent by controlling the post-harvest environmental conditions of temperature, relative humidity and atmospheric concentration of certain gases. There are various techniques and methods to store fruits and vegetables safely and are discussed below:

  - Refrigeration

    The principle of refrigeration is to take out heat. In mechanical refrigeration, the refrigerated gas (eg. ammonia, freon, etc.) takes out the heat from the chamber or store as it expands. The expanded gas is then compressed, and the heat is removed from the compressed air over the tubes containing the hot gases. The gas is liquefied and the cycle is then repeated. With such a system accurate temperature control is possible.. Fruits and vegetables can be safely kept for a long period through refrigeration. However, many vegetables are of low unit value to bear the cost and are also susceptible to low temperature injury.

- Cold storage

In cold store, the temperature control is very important. An ideal environment condition for fresh fruits and vegetables in storage is the lowest temperature, which does not cause chilling injury to the produce. Any variation from the desired condition is detrimental. Relative humidity of the store rooms, also has a considerable bearing on the keeping quality of the horticultural produce. It is very difficult to control the moisture in air. A small temperature difference between the cooling coil and the stored products is required to maintain an adequate control of RH in storage room. In order to increase the RH, water may be sprayed into the controlled storage chamber.

The rate of respiration has direct correlation with temperature. Higher the temperature, more is the rate of respiration and multiplication of decay organisms. The temperature and relative humidity requirements are specific for each fruits and vegetables. The optimum cold storage conditions for fruits and vegetables are given separately in tables. Most of the vegetables require a storage temperature of few degrees above the freezing points excepting okra, cucumber, brinjal, green beans, sweet potatoes, squash and tomato. Low but non- freezing temperatures may lead to chilling injury as in tomato, brinjal etc. and becomes noticeable only after the vegetables are transferred to room temperature i.e. 22° C. A low cost cooling chamber in which a drop in temperature up to 15° C can be produced by evaporative cooling was developed at Indian Agricultural Research Institute (JARI), New Delhi. The technique requires bricks, riverbed sand, bamboo, Khas-khas or any other similar plant material and gunny bags. It consists of a brick-wall of two layers with a layer of sand in between. Careful sprinkling of water on the wall twice a day can bring down the temperature and increase the storage life of different vegetables and fruits from 3 to 15 days. This technology is useful in places where transportation is problem, to reach the market place daily.

Table: Recommended storage temperature and relative humidity for different fruits

Sl.No	Name of the fruit	Storage temp.(%)	Relative humidity(%)	Storage period (weeks)
1	Pineapple	8-10	85-90	1-2
2	Pomegranate	0-1.66	85-90	16-17
3	Guava	8.30-10.00	80-85	4
4	Mango	7.20-8.80	85-90	4-7
5	Emblica (amla)	0-1.66	85-90	7-8
6	Grapes	0-1.66	80-85	6-8
7	Fig			
	(a) Fresh fruits	0-1.66	80-90	4
	(b) Dried fruits	0-1.66	65-70	52
8	Cashewnut	0-1.66	85-90	4-5
9	Jackfruit	11.10-12.70	85-90	6
10	Banana			
	(a) For ripening	15.50-21.00	80-85	1-2
	(b) Ripened fruit	11.10-12.70	85-90	3
11	Date palm			

	(a) Fresh fruits	7.2-8.80	85-90	2
	(b) Dried fruits	0-1.66	65-70	40-52
12	Grapefruits	7.20-8.80	85-90	12
13	Roughlime	5.50-7.20	85-90	13-17
14	Sapota (cheeku)	1.66-3.30	85-90	6-8
15	Cherry	0-1.66	85-90	2
16	Pear	0-1	85-90	13-26
17	Papaya	8.30-10.00	80-85	1-2
18	Passionfruit	5.50-7.20	80-85	4-5
19	Malta			
	(a) Malta common	3.90-5.50	85-90	17
	(b) Blood Red	2.20-3.90	85-90	17
	(c) Mosambi	5.50-7.20	85-90	21
	(d) Valentia late	3.90-5.50	85-90	17
	(e) Sathgudi	5.50-7.20	85-90	17
20	Lime	8.30-10.00	85-90	6-8
21	Litchi	0-1.66	85-90	10
22	Lemon	7.20-8.80	85-90	8-12
23	Strawberry	0-1.66	85-90	5-6
24	Santara	3.90-5.50	85-90	10-14
25	Apple	0-1.66	85-90	17-34
26	Plum	0-1.66	85-90	2-4
27	Peach	0-1.66	85-90	2
28	Bael	8-9	85-90	10-12

Table: Recommended storage temperature and relative humidity for different vegetables.

Sl.No	Name of vegetable	Temperature (0° C)	Relative humidity (%)	Storage life (weeks)
1	Asparagus	0-0	95	3-5
2	Brinjal	10.0-11.10	92	2-3
3	Dolichos lablab (pod)	0.0-1.7	90	3
4	Beet (topped)	0.0-1.7	90-95	8-14
5	Beet (bunched)	0.0	90	1.5
6	Bitter gourd	0.6-1.7	85-90	4
7	Cabbage (early)	0.0-1.7	92-95	4-6
8	Cabbage (late)	0.0-1.7	92-95	12
9	Carrot (topped)	0.0	95	20-24
10	Cauliflower	0.0-1.7	85-95	7
11	Celery	0.6-0.0	92-95	8
12	Colocasia	11.1-12.8	85-90	21
13	Coriander	0.0-1.7	90	5
14	Cucumber	10-11.7	92	2
15	Garlic	0.0	65	28-36

16	Ginger	7.2-10.0	75	16-24
17	Lettuce (head)	0.0	90-95	3
18	Lettuce (leaf)	0.0	95	1
19	Limabean (pod)	4.4-7.2	90-95	1.5-2
20	Muskmelon			
	(a) Cantaloupe	1.7-3.3	85-90	1.5
	(b) Honey dew	7.2	85	4.5
21	Okra	8.9	90	2
22	Onion (leaf)	0.0	90-95	2
23	Onion (bulbs)	0.0	70-75	20-24
24	Pea (green)	0.0	88-92	2-3
25	Pepper (ripe)	5.6-7.2	90-95	2
26	Potato (Irish)	3.0-4.4	85	34
27	Pumpkin	1.7-11.6	70-75	24-36
28	Radish (topped)	0.0	88-92	3-5
29	Squash (winter)	12.8-15.6	70-75	24-36
30	Sweet potato	10-12.8	80-90	13-20
31	Cassava	0-1.7	85	23
32	Tomato (unripe)	8.9-10.0	85-90	4-5
33	Tomato (ripe)	7.2	90	1
34	Turnip	0.0	90-95	8-16
35	Watermelon	7.2-15.6	80-90	2

- Controlled atmosphere storage

The controlled atmosphere (CA) storage in which high level of carbon dioxide (CO_2) and low level of oxygen (o_2) are maintained, is one of the most significant contribution to the storage technology. When combined with refrigeration, it markedly retards the respiratory activity and may delay softening, yellowing, quality changes, and other deteriorative processes. However, important factor in CA storage is the tolerance or susceptibility of the fruits and vegetables to the injury caused by decreased O_2 and increased CO_2 concentration. The relative tolerance of individual fruits and vegetables varies considerably and is largely dependent on temperature, age, morphology and the anatomy of the horticultural products.

In this system, the produce is held under the atmospheric conditions modified by package, over wrap, box liner or pallet cover. The first requirement of the controlled atmosphere is a gas-tight envelope around the produce and the second requirement is some means of maintaining the concentration of CO_2 and O_2 at a desired level.

The fruit that has derived the most benefit from CA storage is apple. Among the tropical fruits, the best atmosphere for storage of mangoes is 5 % CO_2 and 5 % O2 at 13° C.

The beneficial effect of CA storage are (1) prolonged storage life of perishables by arresting the respiration and senescence process, (2) reduction in sensitivity to ethylene, and (3) decrease in the incidence and severity of disease causing organisms such as fungi, bacteria, insects and pests, in selected commodities. The CA storage proves beneficial for fruits and vegetables that deteriorate rapidly or those that complete ripening after harvest.

Although CA storage is beneficial for extending the storage life of fruits and vegetables, it can also cause ad- verse effects such as (1) aggravation of physiological disorders, (2) irregular ripening, (3) development of off-flavour, and (4) accumulation of organic acids. Injury to the tissue or unfavorable changes due to the accumulation of high CO_2 and low O_2 concentrations. No gas mixture is suitable under all circumstances because the temperature, duration of storage and the commodity are interdependent factors, each influencing the use of gas mixture in the CA storage. Although CA storage is a practical method, the probable effects of its use still need critical studies to determine the limits of CO_2 and O_2- The response of citrus fruits to CA storage has been disappointing. CA storage of Florida citrus is not recommended. Fruit decay from stem end rot is a limiting factor and in many case it seems to be increased by high level CO_2, There is also a hazard of ethylene accumulation, therefore, commercial CA storage recommendation cannot be given to citrus fruits.

- Ionizing radiation

Irradiation of food is one of the useful methods developed with the last three decades. One of the significant advantages of the ionizing radiation is that it can destroy the pathogenic organisms without appreciably raising the temperature of the produce. Though it involves high initial capital involvement. The radurization (Pasteurization with ionizing radiations) is being well accepted due to its easy application. it leaves no potential residue and needs minimal energy expenditure. Ionizing radiation extends the storage life of perishables through its action as surface pasteurization, sprout inhibition and retardation of senescence process.

Ripening of mangoes can be delayed by the application of ionizing radiation. Irradiation with 25-35 krads is also useful to delay the initiation of natural ripening of banana without interfering with ethylene induced ripening or affecting the quality of the fruit. Ionizing radiation also inhibits the sprouting of potatoes, onions and other bulbs most effectively at. 10-15 krads.

- Waxing

The outer surface of fruits and vegetables (excluding under-ground ones) have a natural waxy layer, which is partly removed during handling and washing. An extra discontinuous layer of wax applied artificially, provides the necessary protection against the entry of decay organisms. The practical benefit from wax coating is usually a reduction in evaporation and respiration. It has been reported that where refrigerated storage facilities are not available, protective skin coating with wax is one of the methods developed for increasing the storage life of fruits and vegetables.

Fruits and vegetables passed through the water emulsified with mustard oil increase the storage life of many fruits and vegetables like tomato, okra, parwal, mangoes, etc.

- Chemical treatment

There are certain chemicals other than fungicides which increase the storage life of fruits and vegetables by delaying the ripening and senescence process. Post-harvest treatment with gibberellic acid markedly delayed the ripening of banana. Potassium per magnate is used as an ethylene absorbent to keep the ethylene below the threshold level. A commercial preparation of this absorbent called "Purafil" is available. The effect of maleic hydrazide on the ripening process, varies with different fruits. Maleic hydrazide at 1000-2000 ppm (parts per I million) delays the ripening of mangoes while, it may hasten ripening in sapota. Foliar sprays of IPC (isopropyl-N- phenyl carbonate) reduces the rate of respiration, ripening and spoilage of fruits during storage.

Post-harvest applications of plant hormones have been found to prolong the storage life of vegetables .too. The effect of cytokinin is more conspicuous, especially in leafy vegetables. Post-harvest application of gibberelins markedly retards the ripening of tomatoes. To maintain the green colour of pods of different vegetables, they can be treated with Indole acetic acid (1AA) solution. Application of 2, 4-D if combined with BA (M6-benzyladenine), will retard the yellowing of cauliflower and some other green vegetables. Pre-harvest sprays of maleic hydrazide (MH) js used to inhibit onion and potato sprouting in storage. Inhibition of the synthesis of solanine in potato by Alar (M, M-dimethyl amino-succinamic acid) and prolonged storage life of leaf lettuce, have also been reported.

- Use of in-package desiccants

  Recently, to maintain the succulent nature and tender- ness of fruits and vegetables, partial dehydration is done and then in-package desiccants are used to limit the moisture availability to the micro-organisms. The use of in- package desiccants such as calcium oxide or fumed silicas, to bring the moisture content of the storage environment to I % or lower, permits the storage of dehydrated fruits and vegetables for a longer period.

- Use of fungicides

  The control of micro-organisms is of prime importance. Pre-harvest spraying of fungicides can result in significant reduction of the pathogens in the harvested produce. The best method of application of fungicides and antibiotics is to mix them in wax emulsions. Fumigation of stored materials with sulphur dioxide (SO_2) can prevent the spread of fungi like Botrytis. Other fumigants like ethylene oxide, propylene oxide, bi-phenyl, acetaldehyde, nitrogen trichloride can also be used. Biphenyl impregnated tissue wrappers control the growth of Penicilium fungi. Benomyl is also good in preventing the growth of most of the damaging fungi of tomatoes. Bacterial soft rot of packed spinach and lettuce could be prevented by dipping in 1000 ppm streptomycin (50-500 ppm). It has also shown marked inhibitory activity towards a number of micro-organisms, prolong shelf life and added nutritive value to foods.

  As far as possible, post-harvest treatment of fruits and vegetables with poisonous chemicals should be avoided. In case, it is absolutely imperative, the consumption of fruits and vegetables should be withheld at least for one or two weeks as they may create different respiratory and digestive disorders in consumers.

Vegetable Pathogens

Soilborne plant pathogens can significantly reduce yield and quality in vegetable crops. These pathogens are particularly challenging because they often survive in soil for many years and each vegetable crop may be susceptible to several species. Simultaneous infections from multiple pathogens sometimes result in a disease complex that can further damage the crop. Many diseases caused by pathogens are difficult to predict, detect, and diagnose. In addition to this, the soil environment is extremely complex, making it a challenge to understand all aspects of diseases caused by pathogens.

Special terms are used when discussing pathogens. Pathogens are the biological agents that cause or incite the problem. Symptoms are the visible reactions (e.g., root decay, tissue discoloration, crown rot, wilting of foliage, etc.) of the plant when it is infected and colonized by the pathogen. The collective manifestation of symptoms caused by the pathogens is the disease. In contrast to a symptom, a sign is the visible evidence of the presence of the pathogen itself (e.g., mycelial growth, large sclerotia, bacterial ooze, or nematode cysts). Inoculum is the biological object (e.g., spore, mycelium, sclerotium, cells) that is able to infect the host and cause the disease. The term pathogens, therefore, can be defined as pathogens that cause plant diseases via inoculum that comes to the plant by way of the soil.

The most familiar diseases caused by pathogens are probably rots that affect belowground tissues (including seed decay, damping-off of seedlings, and root and crown rots) and vascular wilts initiated through root infections. A few pathogens, however, cause foliar diseases with symptoms and damage appearing on aboveground parts of plants. For example, the lettuce anthracnose pathogen survives in the soil in the form of tiny resting structures (microsclerotia). When raindrops splash pathogen-laden soil particles onto lettuce leaves, the fungus moves onto the plants and causes a leaf spot disease. Similarly, the pathogen Sclerotinia sclerotiorum survives in soil as sclerotia. Under certain environmental conditions, the sclerotia produce tiny mushroom like structures (apothecia) that release aerial spores; these spores then land on susceptible foliage and cause a foliar disease.

Because soil ecology is so complex, it is also important that we define the ecological roles of pathogens. pathogens can generally be divided into soil inhabitants (those able to survive in soil for a relatively long time) and soil invaders or soil transients (those only able to survive in soil for a relatively short time). Many plant pathogens also can function and live as non-pathogenic soil organisms under certain conditions. If these pathogens are in contact with dead and decaying plant tissues, they can grow and survive on these substrates and thus be seen as saprobes or saprophytes (organisms that live on decaying organic matter).

Major Pathogen Groups

The agents that cause diseases make up a diverse group. Fungi, which are multicellular microorganisms, cause most vegetable diseases and so are considered the most important pathogen group. Plant-pathogenic fungi fall into five main taxonomic classes based on morphological and biological characteristics: Plasmodiophoromycetes, Zygomycetes, Oomycetes, Ascomycetes, and Basidiomycetes. Some species of Ascomycetes and Basidiomycetes form a second type of spore that is asexually produced. These asexual stages are placed in an additional, separate class, the Fungi Imperfecti. Notable Oomycetes pathogens include Aphanomyces, Bremia, Phytophthora, and Pythium. Important Ascomycetes are Monosporascus and Sclerotinia. Examples of Fungi Imperfecti pathogens are Fusarium, Rhizoctonia, and Verticillium. Plasmodiophora brassicae (causal agent of clubroot disease of brassicas) and Spongospora subterranea (causal agent of powdery scab of potato) are the main Plasmodiophoromycetes pathogens. Many fungi persist in soil for long periods because these organisms produce resilient survival structures like melanized mycelium, chlamydospores, oospores, and sclerotia. The thin-walled mycelium typical of many fungi survives for only a short time in the soil.

Bacteria are single-celled organisms that have rigid cell walls but lack a membrane-bound nucleus. Fewer diseases are caused by bacterial pathogens than by fungal pathogens. Examples of such bacteria are Erwinia, Rhizomonas, and Streptomyces. Pathogens in the Pseudomonas and Xanthomonas groups usually persist in the soil for only a short time.

There are few viruses that affect vegetable crops. Viruses are subcellular entities composed of genetic material with a surrounding protein coat. Once the genetic material of a plant virus is inserted into a host cell, it causes the cells to manufacture more virus particles. Virus disease symptoms include stunting of the plant, tissue distortions, and discolorations of foliage and fruit. viruses generally survive only in the living tissues of the host plant or in the nematode or fungal vectors that transmit them to the plant hosts. Lettuce big vein disease, for instance, is caused by the Mirafiori lettuce virus; the virus is present inside a primitive soil fungus (Olpidium brassicae) that moves in soil water, attaches itself to lettuce roots, and transmits the virus. Another example of a virus that affects vegetables is the lettuce necrotic stunt virus (LNSV). This is an unusual case in that LNSV has no known biological vector, but is found in river water and in soils contaminated by such water.

Nematodes are tiny, non segmented roundworms. plant-parasitic nematodes spend most of their lives in the soil, either as external feeders on plant roots or as residents inside roots. Nematodes affect crops by reducing plant vigor and growth. In an affected field, some plants will be heavily infested and others will not, with the result that the overall crop will mature unevenly or the quality of the produce will be lower. In soil, plant parasitic nematodes either live freely or are present as eggs or durable cysts. Root knot nematodes (Meloidogyne species) cause a general reduction in vigor for many plant species and can cause severe distortions and swellings of roots, particularly affecting the marketability of root crops such as carrots. Cyst nematodes (Heterodera species) can survive in the soil for long periods because the mature body of the female cyst nematode dries in the form of a leathery cyst, protecting the eggs within. Needle nematodes (Longidorus africanus) feed on the growing points at the tips of the roots, causing root tips to swell and causing roots to fork or branch out. Stubby root nematodes (Paratrichodorus species) reduce the length of roots.

Common bacterial diseases and crops affected:

Some examples of common bacterial diseases of vegetable crops are provided in the table below with some typical symptoms.

Bacterial disease	Factors conducive to spread	Crops affected	Symptoms
Black rot (Xanthomonas campestris pv. campestris)	Warm, wet conditions.	Brassicas.	Light-brown to yellow V-shaped lesions on the leaf, which become brittle and dry with age. Vein blackening with the necrotic area.

Bacterial canker (Clavibacter michiganensis pv. michiganensis)	Moderate temperatures and high humidity.	Tomato; capsicum; chilli	Seedlings may die and older plants may wilt and die eventually. Older plants have leaves that turn yellow and wilt only on one side. Cankers on stems and fruit. Tissue inside stems becomes discoloured.
Bacterial soft rot (Pseudomonas spp., Erwinia spp.)	Warm, wet conditions.	Wide range of vegetables, including lettuce; brassicas; cucurbits; tomato; capsicum; potato; sweetpotato; carrots;herbs.	Wet, slimy, soft rot that affects any part of vegetable crops including heads, curds, edible roots, stems and leaves. May have a disagreeable odour.
Bacterial leaf spot/ Bacterial spot (Xanthomonas campestris - various strains)	Overhead irrigation and windy conditions.	Range of vegetables including lettuce; cucurbits; tomato; capsicum.	Lettuce – Large brown to black circular areas that start as small translucent spots; usually on outer leaves. Tomatoes and capsicums – Greasy spots on leaves and stems that go from tan to black; fruit may have circular spots with central scab. Cucurbits – Begin as small water-soaked/greasy spots on underside of leaves with corresponding yellowing on upper side; fruit may produce light-brown ooze from water-soaked markings.
Bacterial wilt (Ralstonia solanacearum)	High temperatures, high soil moisture and poor drainage. Once infection has occurred, severity of symptoms is increased with hot and dry conditions, which facilitate wilting.	Potato; tomato; capsicum; eggplant.	Wilting, yellowing and stunting of plants but they may wilt rapidly and die without any spotting or yellowing; vascular tissue appears brown and water-soaked; a white ooze appears when pressure is applied to affected tubers or stems.
Bacterial leaf spot/ Bacterial spot/Bacterial blight (Pseudomonas syringae - various strains)	Long periods of leaf wetness.	Beet; spring onions; leeks; rocket; coriander.	Beet – irregular, round leaf spots with a grey centre surrounded by a purple margin. Spring onions/shallots – pale yellow to light-brown lesions with a water-soaked appearance around the margins; outer leaves wither and die and youngest leaf turns lemon to light-green. Leeks – brown streaking on the shank.
Bacterial blight (Pseudomonas syringae pv. pisi)	Cool, wet, windy conditions.	Peas.	Water-soaked spots on leaves and stipules which become dark-brown and papery in warm weather or black in cool weather. Water-soaked spots on pods that become sunken and dark-brown.

Bacterial speck (Pseudomonas syringae pv. tomato)	Humidity and overhead irrigation.	Tomato.	Small dark spots surrounded by a yellow halo on leaves; dark raised specks on fruit.
Bacterial brown spot (Pseudomonas syringae pv. syringae)	Cool, wet, windy conditions.	Beans.	Tan to reddish-brown spots on leaves. Water-soaked spots on pods which enlarge and become sunken and tan with distinctive reddish-brown margins.

Other bacterial diseases of vegetables include:

- Peppery leaf spot – *Pseudomonas syringae pv. maculicola* (brassicas);
- Varnish spot – *Pseudomonas spp.* (lettuce);
- Corky root – *Rhizomonas suberifaciens* (lettuce);
- Angular leaf spot – *P. syringae pv. lachrymans* (cucurbits);
- Bacterial pith necrosis – *Pseudomonas corrugata* and other bacteria (tomatoes);
- Common bacterial blight – *Xanthomonas campestris pv. phaseoli* (beans);
- Halo blight – *Pseudomonas syringae pv. phaseolicola* (beans);
- Black leg – *Erwinia carotovora pv. atroseptica* (potatoes).

Management

Disease management strategies aim to favour the host plant's growth and development while attacking vulnerable stages in the lifecycle of the pathogen to prevent or restrict its development. The key means of bacterial disease management include:

- Exclusion or eradication of the pathogen (quarantine and use of pathogen-tested seeds and propagated materials);
- Use of clean transplants;
- Monitor crops regularly and use predictive models;
- Reduce the pathogen levels by crop rotation;
- Remove weeds and incorporate crop residues that can host the disease;
- Protect the host plant by using resistant plant varieties;
- Minimise mechanical damage to crops and damage by insect pests;
- Avoid working in crops when they are wet;
- Spray with a registered bactericide when weather conditions favour disease development to prevent infection;

- Understand chemical resistance and rotation of chemical groups;
- If the plants are already infected, isolate and destroy them and prune infected leaves, but avoid excessive handling of diseased plants; if the disease is systemic and has spread throughout the plant, the plant cannot recover and should be destroyed (burning or burying);
- Use correct temperatures and packing conditions during transport and storage.

Fruit Pathogens

Pathogens which adversely infect fruits are known as fruit pathogens.

Fruits can be contaminated with microbes from the soil, the air, water or animal wastes. The high water content and acidity (low pH) favour the growth of fungi – yeasts and moulds.

Whilst still on the plant, fruits can be infected by pathogenic microbes which cause wilts, blotching and browning. Once harvested care has to be taken in handling as bruising can allow the entry of harmful organisms, particularly fungi, which soon rot the product. The green mould sometimes seen on oranges is a type of Penicillium; Botrytis causes the fuzzy grey growth on strawberries.

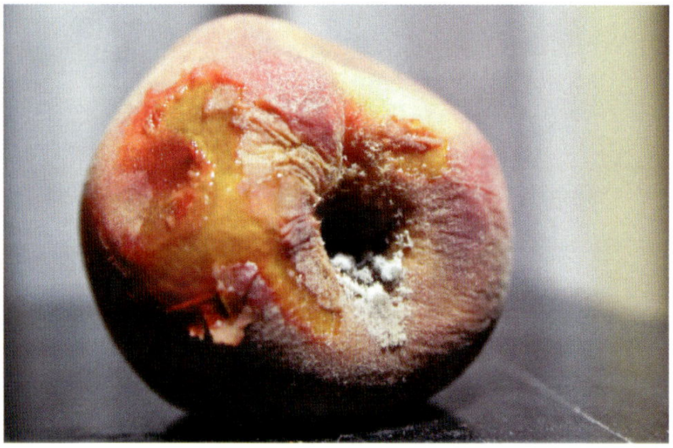

A mouldy peach.

Washing, dipping or wrapping fruit in paper that has been impregnated with preservative is sometimes used to prolong storage life. Refrigeration and controlled atmosphere packaging also allow fruit to be kept for several months before being sold – this is how fruit from all around the world can appear on our supermarket shelves.

Dishes such as fresh fruit salad contain sugar syrup. If not eaten quickly, yeasts from the air soon get to work and ferment it.

Food poisoning is rarely caused by eating fruit, although pathogenic bacteria can sometimes get on them from the soil and manure. Some intestinal virus diseases have been spread on fruit due to contamination from human faeces either during or after harvesting. Washing and good hygienic practices should prevent this.

Mould growth can produce poisons known as aflatoxins in certain fruits. For example *Penicillium expansum* and other fungi on apples cause the formation of patulin which can get into the fresh juice. It is not a good idea to eat mouldy fruit.

Postharvest Losses in Grains

A variety of factors can lead to a loss in agricultural productivity of grains. This loss may occur at all stages of production, and includes pre-harvest, harvest and post-harvest stages. In case of the pre-harvest stage, the contributing factors to agricultural loss are weeds, insects and rusts. Losses between the beginning and completion of harvest mainly occur as a result of shattering. From harvest to human consumption, grain loss can occur when the grain is winnowed, threshes and dried, as well as when it is stored, transported or processed.

Rice, wheat, and maize are major cereal grains in most of the developing countries. In countries such as Bangladesh, rice accounts for more than 90% of food produced and about 70% of calories intake. In West Africa, Nigeria is currently the largest producer of rice with an annual production of about 3.3 million tonnes. In spite of the large production and huge rice imports every year, a large number of people are undernourished in Nigeria. Similarly, Bangladesh is the fourth largest producer of rice worldwide, however, it is still food deficient and imports more than one million tons of rice every year. Saving the cereal crop lost during postharvest operations can help in meeting the food demand and reduce the load on the economy. A report from the World Bank, estimated 7%–10% of grain loss in postharvest operations at field level, and 4%–5% loss at the market and distribution stage in India for the year 1999. Estimates also suggest that these, approximately 12 to 16 million metric tons of grains wasted each year, could meet the food demand of about one-third of India's poor population. However, despite the criticality of the issue, availability of consistent and reliable postharvest loss data is still a challenge. Very few loss assessment studies have been conducted in important developing economies such as India, China, and Brazil. After the FAO report in 2010, various institutes are making efforts to conduct comprehensive household surveys, interviews, and field measurements to determine the actual status of losses along food supply chains in various countries.

Rice

Being a high energy calorie food, rice accounts for one-fifth of the global calorie supply. The scale of postharvest losses in grain supply chains varies significantly, depending on the economy, agricultural conditions and practices, and climatic conditions of the region. An example of such variation of losses in the rice supply chain in different countries is illustrated in figure. The rice losses have been reported as low as 3.51% in India to as high as 24.9% in Nigeria. Based on the 24.9% loss, the value of the total grain loss during the rice supply chain in Nigeria was estimated as 56.7 billion Nigerian Naira (NGN). According to Bala, the losses in the rice value chain from producer to retailer were estimated as 10.74% to 11.71% (10.74% for Aman, 11.71% for Boro, and 11.59% for Aus) in Bangladesh. Most of the losses (85.28%–87.77% of the total) happened in the farm level operations, with storage losses (33.92%–40.99% of farm level losses) being the main contributor. Alavi et al. compiled data on postharvest losses in rice value chains from different studies conducted by the FAO and reported 10%–37% losses in rice in Southeast Asia. In China, the losses were estimated in the range of 8%–26%.

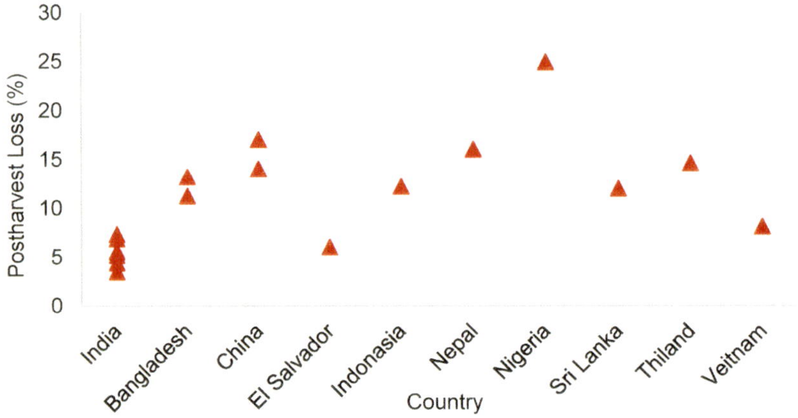

Figure: Postharvest losses in the rice value chain in various countries (in the case of a range of losses, an average of losses was used).

Wheat

Wheat is another major staple food of several countries in Europe, Asia, and North America. Similar to rice, significant losses happen during postharvest processing of wheat in developing countries. According to data compiled (before the year 1978) by US National Academy of Sciences, wheat losses in Sudan and Zimbabwe were estimated 6%–19% and 10% respectively. Bala et al. reported that the storage losses were maximum (41.7% of the total) among all the postharvest operation losses for wheat in Bangladesh, even considering the fact that the storage period of wheat is relatively small. Basavarja conducted a study to estimate losses in postharvest operations of wheat in the state of Karnataka, India. The estimations were based on a comprehensive survey from 100 farmers, 20 wholesalers, 20 processors and 20 retailers from the major producer district of each crop in the Karnataka state. The overall losses in the wheat supply chain from harvesting to retailer were estimated as 4.32%. Field level operations contributed 75.9% among the total postharvest losses. The losses were maximum during storage operations due to a lack of availability, poor structures, presence of rodents, and improper drainage.

Table: Postharvest losses of wheat from various studies in different countries.

Country	Year	Losses (%)	Comments
Bangladesh	2010	3.62	Maximum losses during storage
India	2013	1.84	Maximum losses during harvesting Punjab (study locality) is a developed state, so has better storage practices
	2013	2.74	Study conducted in Uttar Pradesh Maximum losses during harvesting (58.4% of the total)
	2004	4.32	Study conducted in Karnataka Maximum losses (0.95%) during storage at field level
	2012	4.32	About 75% losses at the farm level Maximum losses during storage at field level (28.9% of the total loss)

India	2012	8.61	Study conducted in Madhya Pardesh Maximum losses during storage (56% of the total losses)
	2013	7.22	Study conducted in West Bengal Maximum losses during storage (54% of the total)
	2013	11.71	Study conducted in Assam Maximum losses during threshing (28.3%) and transportation (25.2% of the total)
Peru	2012	15–25	-
Sub-Saharan Africa	2013	15.2	-

Maize

Maize is an important part of staple food in Sub-Saharan Africa (SSA) and a major source (~36%) of daily calories intake. Pantenius estimated 0.2%–11.8% weight loss due to insect infestation in maize after 6 months of storage in traditional granaries in Togo. Inter-American Institute for Cooperation on Agriculture (IICA) conducted a survey to estimate the postharvest losses in Latin America and Caribbean countries. The losses values estimated in various regions have been presented in the table below. In almost all regions, most of the losses were observed occurring at the small and medium-scale farms due to a lack of adequate harvesting, drying, and storage technologies, along with a lack of information about the good agricultural practices. In Guatemala, due to a lack in storage structures along with the region's high humidity, storage losses were estimated between 40% and 45%. Insect infestation was found as the major reason of storage losses in most of the cases. Kaminski and Christiansen conducted a study to estimate the postharvest losses in maize crop in three SSA countries (Uganda, Tanzania, and Malawi) through comprehensive household surveys. The losses from the farm level activities were estimated in the range of 1.4% to 5.9%. Insects and pests were reported as the major reason of losses in maize during storage. Alavi et al. reported an average of 23% losses in the maize value chain in ASEAN (the Association of Southeast Asian Nations) countries, with maximum losses happening during field drying (9%). Most of the maize is dried along the sides of road, especially in the Philippines. In Vietnam, major losses occur due to rodent attack and fungal disease during maize storage.

Table: Postharvest losses of maize from various studies in different countries.

Country	Year	Losses (%)	Comments
Bangladesh	2010	4.07	Maximum losses during storage (60.4% of the total)
Ecuador	2012	10–30	Major losses due to insect infestation during storage
Guatemala	2012	50	-
Malawai	2010	1.4	These losses are only from farm level activities
Panama	2012	20	Major losses at the small-scale level farm due to a lack of adequate technology
Peru	2012	15–25	-
Sub-Saharan Africa	2013	17.8	-

Country	Year	Losses (%)	Comments
Tanzania	2008	4.4	These losses are only from farm level activities
	2010	2.9	
Uganda	2009	5.9	These losses are only from farm level activities

The maximum amount of losses occurs during the storage of crops due to a lack of adequate infrastructure. Storage losses can be classified in two categories: direct losses, due to physical loss of commodities; and indirect losses, due to loss in quality and nutrition. It is important to consider both damage and losses by the insects during storage instead of just weight loss. "Damage" can refer to physical evidence of deterioration, for example, holes in the grains. It mainly affects the quality of grains. "Loss", on the other side, is the total disappearance of the food, which can be measured quantitatively. The loss in quality results in value loss of the product, and sometimes leads to total rejection also. The rejection rate depends upon the individual's economic status and cultural background. For example, a subsistence farmer may consume the damaged food to some extent, whereas, affluent customers may reject even slightly damaged food. Some loss happens in the form of spillage from leaky sacks which can be identified when the store is emptied and the spilt grain remains on the floor.

The storage losses are affected by several factors, which can be classified into two main categories: biotic factors (insect, pest, rodents, fungi) and abiotic factors (temperature, humidity, rain). Moisture content and temperature are the most crucial factors affecting the storage life. Most of the storage molds grow rapidly at temperatures of 20–40° C and relative humidity of more than 70%. Low moisture keeps the relative humidity levels below 70% and limits the mold growth. In the traditional storage structure, temperature fluctuations due to weather changes cause moisture accumulation either at the top or bottom of the grains' bulk depending on the direction of air convection. This can be avoided by minimizing the temperature difference of inside and outside the storage structure. Grains should be dried to about 13% of the moisture content before storage to minimize the losses. At moisture contents of 16% or higher, the safe storage period of rice is only a few weeks. Quality of grains before storage is another critical factor affecting the storage losses. Mechanical damage during harvesting and threshing can result in bruised areas on grains, which may serve as centers for infection and cause deterioration. The criticality of a factor depends upon the storage conditions.

In most of the developing countries, especially in Africa and South Asia, grains are generally stored as bulk or bags in simple granaries constructed from locally available materials (straw, bamboo, mud, bricks). Mud bins and pots, bokharies (straw structure), kothis, and plastic containers are common storage structures in Asia . Gunny bags and Plastic/Polythene bags are commonly used for the short duration storage, and Dole, Berh, Gola, Motka, Steel/Plastic drums are used for the long duration storage. Various types of granaries are used in African countries. Ebli-va, Kedelin, in-house smoked storage are some of the common maize storage structures in Togo. In the in-house smoked storage methods, maize is stored within the dwelling in space between the ceiling and roof over the cooking spot to receive the heat. In West Africa, grains are commonly stored in the home or field in jute or polypropylene bags, raised platforms, conical structures, and baskets. In East and Southern Africa, farmers use cow dung ash in small bags, wood cribs, pits, iron drums enclosed with mud, and metal bins for storing the grains. "Nkokwe" is one of the most commonly used storage structure used in Malawi and Kenya. This is a kind of cylindrical basket made up of interwoven split bamboo and covered with a conical shaped roof of grass. The structure is raised off the ground on stilts. Most of these structures are not scientifically designed and are made from locally available materials, and cause damage to stored grains due to biological, environmental and other factors.

Insect Infestation

Among all the biotic factors, insect pests are considered most important and cause huge losses in the grains (30%–40%). Some studies in Ghana reported that the maize losses due to insect infestation could be up to 50%, if all quantity losses, quality losses, and income loss due to early sale are considered in the estimation. According to an economic model by Compton et al. each percent of insect infestation results in 0.6%–1% depreciation in the value of maize. From field studies in Togo, Pantenius observed that insects and pests were responsible for 80%–90% of storage losses in grains. *Callosobruchus maculatus* (F.) alone, a common pulse weevil has been found responsible for up to 24% losses in stored pulses in Nigeria. Losses due to insects in stored maize have been reported from 12% to 44% in the western highlands of Cameroon. The maize weevil (*Sitophilus zeamais*), and larger grain borer (LGB) (*Prostephanus truncatus*) are the major pests in the maize. About 23% losses were observed in maize grains stored for six months, mainly due to infestation of maize weevil and LGB in Benin. LGB originated in Central America and was accidently introduced in Africa in late 1970s. Nowadays, it is found in most parts of Africa and is considered the most threatening pest, as it causes extensive damage in a very short time. The sporadic nature of LGB makes even its control difficult: it does not infest all stores of the same area, and its reoccurrence in each year is not guaranteed. At farm level storage, more than 30% of weight loss have been observed in maize due to these pests. Some studies on maize losses in Ghana estimated about 5% to 10% loss in market value due to infestation by only *Sitophuilus* spp., and 15% to 45% market value loss due to damage by LGB. Overall, these losses were equivalent to about 5% of the average household income in that area. Abass et al. reported that after six months of maize storage, LGB was responsible for more than half (56.7%) of the storage losses, followed by losses due to grain weevil and lesser grain borer. Patel et al. (1993) observed about 25% losses by *R. dominica* in wheat stored for 3 months under laboratory conditions.

Mycotoxins

Mycotoxin contamination is another big challenge, especially in the case of maize, which makes the food unsuitable for human consumption or animal feed. A large amount (25%–40%) of cereal grains are contaminated by the mycotoxins produced by storage fungi world-wide. Molds and mycotoxins cause dry matter as well as quality loss, and are a hazard in the food value chain. Aflatoxins, Fumonisins, Deoxynivalenol, and Ochratoxin are the most common and important mycotoxins, especially in maize. Aflatoxins, produced as secondary metabolite by two fungi species *Aspergillus flavus* and *A. parasiticus*, are considered the most dangerous group of mycotoxins, as they increase the risk of liver cancer and affect growth in young children. Because of food contamination, about 4.5 billion people are exposed to aflatoxins in developing countries. High concentrations of aflatoxin can lead to aflatoxicosis, which can cause severe illness and even death. *Penicillium verrucosum* (ochratoxin), a major mycotoxigenic mold is commonly found in damp cool climates (e.g., Northern Europe) and *Aspergillus flavus* is mostly observed in temperate and tropical conditions.

Mold during storage damages the grains as well as reducing grain germination. It also deteriorates the grain quality due to the must odour, increased fatty acid content, and reduced starch and sugar contents. Lipid peroxidation is another phenomenon that causes food deterioration and alters the taste and aroma, and may cause undesirable effects on human health. High oil content varieties of oil seeds demand particular attention during storage, as the high level of moisture degrades the vegetable oil and produces high fatty acids, which sometimes also results in self-heating. At farm level storage in developing countries, even rodents can damage a large portion of crop, whereas, fungi can be a major reason for spoilage at high relative humidity storage. Use of scientific storage structures and proper handling of grains can reduce storage losses to less than 1%.

Losses can be minimized by physically avoiding the entry of insects and rodents, and maintaining the environmental conditions that avoid growth of microorganisms. The knowledge of control points during harvesting and drying before storage can help in reducing losses during the storage of cereals. Taking the timely preventive actions for biotic and abiotic factors can be very effective in reducing the losses during storage.

Interventions to Reduce Storage Losses for Smallholders

Although a huge challenge, storage losses can be mitigated by use of efficient storage technology, upgrading infrastructure and storage practices. World Food Programme (WFP) with the help of the government and non-governmental organizations (NGOs) performed an Action Research Trial in Uganda and Burkina Faso to demonstrate the impact of improved postharvest management practices and using new storage technologies on the crop loss after harvesting. The results concluded that irrespective of crop or storage periods, use of improved practices and new technologies resulted in about a 98% reduction in food loss. It was also observed that losses in the traditional storage structures were much higher than those reported in the literature, because the storage period was longer than that commonly used by farmers in these countries. It is important to understand their usefulness, technical efficacy, and limitations to promote their adaptability among the consumers. Other than saving losses, the availability of low cost and effective storage structures can motivate farmers to store their grains and obtain high prices instead of selling right after harvesting when there is an abundant supply of grains.

Chemical Fumigation

Synthetic insecticides are used in several countries and play an important role in controlling the pests and reducing losses during storage of grains. Methyl bromide (MB) and phosphine are the most commonly used chemicals in developing countries. If the maize grains are sufficiently dry (moisture content less than 13%), use of phostoxin can control the LGB infestation in maize grains. However, phostoxin can be applied only by the licensed technicians in most parts of Africa, and farmers are allowed to use a mixture of pirimiphos-methyl (Actellic) and permethrin, commercially sold as Actellic Super. Shelling the grains and storing them in polypropylene bags after proper application of Actellic Super can effectively avoid the pest infestation for a few months of storage. This practice has been widely adopted by farmers in African countries, especially Kenya. More than 93% of farmers reported this as the most common method used for controlling pest during storage in Tanzania.

Irrespective of their effectiveness, the synthetic insecticides suffer from limitations such as high costs, development of genetic resistance in the treated pests, health hazards due to toxic residues, and environmental contamination. Residuals from synthetic fumigants could cause considerable loss of seed viability. Due to long use of phosphine, some insects have gained resistance to chemical fumigation in some countries. Use of these chemical fumigation methods is even challenging in the traditional storage structures used in the developing countries, as most of them are open to reinfestation. Another challenge with these chemicals is knowledge and training to apply these pesticides at the correct time and at the correct dose. The delayed treatment, adulterated chemicals, and incorrect dosage can reduce the efficacy of the treatment and result in high storage losses.

Natural Insecticides

Several plant species and their extracts have been found with natural pesticide ability and are used very commonly as a traditional practice to protect the grains from insects in several African and Asian countries. Plant based chemicals and products would be biodegradable, environment friendly, and relatively safe for human health. The leaves and oil extract from leaves of *Chenopodium ambrosioides Linn.* (Chenopodiaceae) has been found to be very effective in controlling the damage of cereal grains by the insects during storage, in several studies. The plant is a branched herb and widely available in India. Its natural pesticide abilities have been highlighted and investigated in several studies. Kumar investigated the effectiveness of essential oil from wormseed against the fungal deterioration in stored wheat. The samples were analyzed for fungi after 12 months of storage at laboratory conditions. The oil was found to be significantly effective in controlling the *A. flavus* fungi in both inoculated (91.17% protection) and uninoculated (99.42% protection) wheat samples. The efficacy of oil was compared with synthetic fungicides (benzimidazole (Benomyl), diphenylamine (DPA), phenylmercuric acetate (Ceresan) and zinc dimethyl dithiocarbamate (Ziram)), and the oil was found to be relatively more effective, with minimum inhibitory concentration (MIC) (concentration at which the oil shows absolute fungitoxicity) lower than those of synthetic ones.

Tapondjou investigated the effectiveness of using leaves and extracted essential oil from wormseed (*Chenopodium ambrosioides* L.) against six common species of grain beetles in the western highlands of Cameroon: *Sitophilus zeamais* (maize weevil)., *S. granarius* (L.) (granary weevil), *Callosobruchus chinensis* (L.), *C. maculatus*, *Acanthocelides obtectus* (all Bruchidae), and *Prostephanus truncatus* (bruchids) (larger grain borer). The ground leaves were mixed with grains at the concentration of 0.05%–0.8% (*w*/*w*) for bruchids, and 0.8%–6.4% (*w*/*w*) for weevils and borers. The grains tested were maize for *S. zeamais* and *P. truncatus*, whole wheat for *S. granarius*, green peas for *C. chinensis*, mung bean for *C. maculatus* and white bean for *A. obtectus*. The oil effectiveness was checked on the filter paper discs by treating a Whatman No. 1 filter paper with the oil diluted in acetone at different concentrations of oil (0 to 1.6 $\mu L/cm^2$). The mortality rate was found to be relatively high for bruchids (100% for *C. chinensis* in 48 h) compared to that in the case of weevils and borers. The essential oil was observed to be more toxic than using leaves and resulted in high mortality. Other than mortality, the ground leaves were effective in inhibiting the F1 progeny production and adult emergence of the insects. Shaaya et al. tested four edible oils: Pure soybean oil, pure and crude cottonseed oils, crude rice bran oil and crude palm kernel oil, as fumigants against common insects in beans and wheat. Crude palm kernel and crude rice bran oils were found to very effectively control *C. maculatus* in chickpea for the initial 4–5 months. Even at the end of 15 months, the insects in the fumigated samples were only 10% of those in the control sample. Similarly, crude cotton and soya bean oils were found to be effective against *S. oryzae* in wheat for the initial 4–5 months, and were effective later at high concentrations (10 g/kg). Only crude cotton oil was found to be significantly effective in controlling *S. zeamais* in maize, and that too mainly during the initial 4 months of storage period. After 8 months of storage, the oil was partially effective and had 20%–40% of the amount of insects of those in untreated maize samples. The major issue with these plant materials is that oil yields are low and might be expensive to use on a commercial scale, however, some plant leaves can be used as natural insecticides by smallholders.

Hermetic Storage

Hermetic storage (HS), also known called as "sealed storage" or "airtight storage", is gaining popularity as a storage method for cereal, pulses, coffee, and cocoa beans in developing countries, due to its effectiveness and avoidance of the use of chemicals and pesticides. The method creates an automatic modified atmosphere of high carbon dioxide concentration using sealed waterproof bags or structures. As the structures are airtight, the biotic portion of the grains (insects and aerobic microorganisms) creates a self-inhibitory atmosphere over time by increasing carbon dioxide concentration (oxygen decreases) due to its respiration metabolism. Some studies have reported that the aflatoxin production ability of *Aspergillus flavus* is also reduced at high concentrations of CO_2. Hermetic storage has been observed to be very effective in avoiding the losses (storage losses less than 1%) during long distance (international) shipments also. Ease of installation, elimination of pesticide use, favorable costs, and modest infrastructure requirements are some of the additional advantages that make the hermetic storage options attractive.

The relationship of the main factors affecting the respiration of grain and microorganisms in hermetic storage has been described in the figure. The CO_2 concentration inside the bags is usually used as an indicator of the biological activity of grains. Permeability of the bag and the gas partial pressure effect the movement of gases (O_2 and CO_2) in and out, whereas the concentration of these gases inside the bag depends on the balance between these exchanges and the respiration of the biotic portion of grains. Higher initial moisture content tends to increase the CO_2 concentration because of the increased respiration, however, the change was not found to be significant.

Another factor affecting the respiration rate is grain temperature. It has been observed in experimental studies that the temperature inside the bags follows the ambient temperature trend, and for every 10-degree increase in temperature, the CO_2 concentration increases by about 1.5% . World Food Programme (WFP) in their Action Research Trial in Uganda and Burkina Faso, found out that if properly sealed, the hermetic storage units were themselves very efficient in killing the pests and insects without any use of phosphine fumigation. Various hermetic storage options, such as Metallic silos, Purdue Improved Cowpea Storage (PICS) bags, SuperGrain bags, etc., have been developed and widely promoted in the last few years. These bags are being considered practical and cost-effective storage technology, and are becoming very popular in several countries .

A metal silo is a strong hermetically sealed structure (mostly cylindrical), built using a galvanized steel sheet, and has been found to be very effective for storing grains for long periods of time and avoiding insects and rodents. In some locations, the siloes are made of painted aluminium sheeting which helps prevent corrosion and improves their appearance. It is considered to be one of the key technologies which will be helpful in reducing postharvest losses and improving food security of smallholder farmers. PICS or Purdue Improved Crop Storage bags, originally developed for storage of cowpea, involve triple bagging the grains in hermetic conditions, and is widely used by farmers in sub-Saharan Africa. The grains are stored in double layer thick (80 μm) high density polyethylene (HDPE) bags and is held in a third woven nylon bag. After filling with the grains, the bags are sealed airtight. This will cut off the oxygen to the weevils and hinder their metabolic pathways preventing them from producing water, and killing them by desiccation. More than 3 million PICS bags were sold in West and Central Africa during 2007–2013. SuperGrain, commercialized by GrainPro Inc. is another widely used water resistant and hermetic storage option. These bags are made up of a single thick layer of high density polypropylene with a thickness of about 78μm, and used as liner along with normal woven polypropylene bags. ZeroFly bags are a product of Vestergaard, Switzerland. These are insecticide infused woven polypropylene bags designed to prevent damaging pest infestations. The bag is made with pyrethroid incorporated into polypropylene yarns.

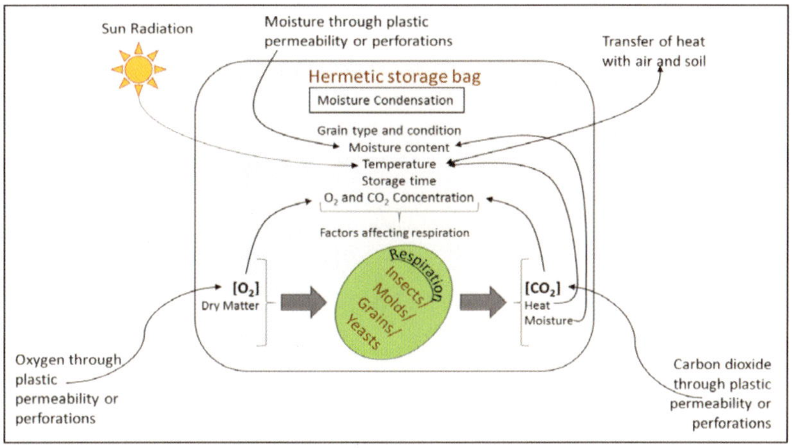

Figure: Illustration of the factors affecting the grain and microorganism respiration in the hermetic storage.

Baoua et al.conducted a study to compare the performance of SuperGrain bags and PICS bags, available in the West African market, to control pest infestation over the 4 months of storage. The change in temperature and relative humidity in both the bags were similar over time. The infestation level of *C. maculatus* eggs on the seeds after four months was found to be lower in SuperGrain bags (18.5 eggs per 100 seeds) in comparison to PICS bag (26.1 eggs per 100 seeds). Grain damage was observed relatively lower (18.5 grains with holes per 100 tested grains) for the PICS bag compared to that in the SuperGrain bags (29.5 grains with holes per 100 tested grains). Somavat et al. compared the effectiveness of hermetic bin bags (GrainSafe IIITM, GrainPro Inc., Concord, MA, USA), metallic bins and gunny bags for storage of wheat under ambient conditions in India. There was no insect infestation found in clean grains stored in hermetic bags after 9 months of storage. For the artificially infested grains, bored grain percentage remained stable at 0.33% for hermetic bags in contrast to 2% and 8% for metallic bins and gunny bags respectively. At end of storage, seed viability was found to be higher (88%) for hermetic bags compared to 82% and 73% in metallic bins and gunny bags respectively.

Baoua et al. conducted a comprehensive study to investigate the effectiveness of PICS bags (50 kg capacity) for maize storage at eleven localities in Burkina Faso, Ghana, and Benin. Insect infestation in maize during storage varied from nil to highly infested. After about 196 days of storage, the PICS bags were able to maintain 100 seed grain weight, seed viability, and seed germination along with 95%–100% insect mortality at all localities. Moisture content was also observed to be unchanged during storage for most of the PICS bags. Although aflatoxin levels were observed in maize stored in both the PICS and woven bags, the level of contamination was lower in the PICs bags. Similar effectiveness of the PICS bags was observed during storage of Bambara groundnut, maize, mung bean and pigeonpeas , pigeonpeas . Mutungi investigated the effectiveness of the PICs bags for mung beans and pigeonpeas. Naturally and artificially infested grains were stored in the PICS bags and woven bags for 6 months. For both mung beans and pigeonpeas, the oxygen levels were reduced and carbon dioxide levels rose rapidly within the two months of storage in the PICS bags. For the initial two months, the change was found to be higher in highly infested grains compared to naturally infested grains, however, at the end of storage the average change was almost the same. Insect damage and weight loss for grains remained unchanged in the PICS bags, whereas, there was 24.2–27.5 times and 21.7–43.7 times more weight loss after 6 months of storage in the woven polypropylene bags for mung beans and pigeonpeas respectively. Treatment of grains with Actellic Super dust before storage in the woven bags did not help in reducing damage, and weight loss at the end of storage increased by factors of 20.8 and 22.5 for mung beans and pigeonpeas respectively.

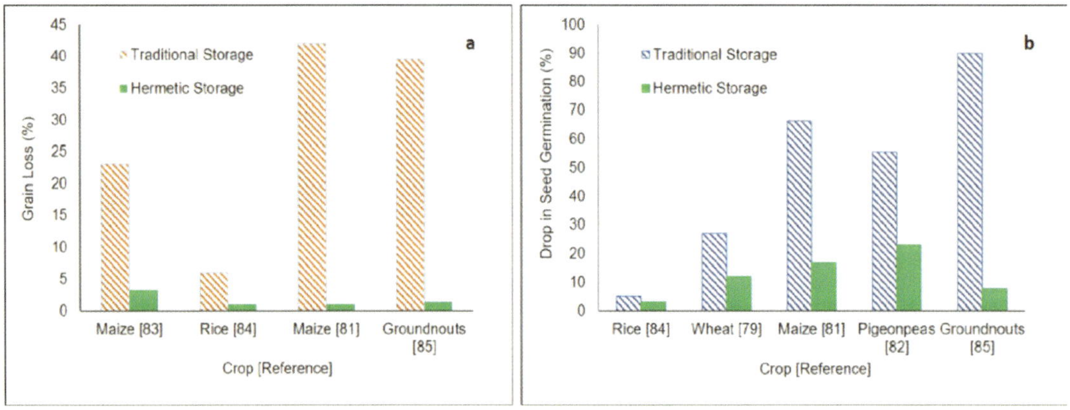

Figure: Amount of losses (a) weight loss; (b) seed germination losses, for various grains due to natural or artificial insect infestation during storage in traditional storage vs. hermetic storage (in the case of a range of losses, an average of the losses was used).

Table: List of the effective use of hermetic bags for various crops in developing countries.

Type of Storage	Crop	Country	Duration of Storage	Investigations	Findings
Super-Grain Bags	Maize	Kenya	6 months	Evaluated performance of hermetic storage (metal silos and super grain bags) and polypropylene bags to control infestation of pests.	Metal silo was the most effective option in controlling pest infestation. Metal silo was equally effective in controlling pest infestation even without any insecticide use. Supergrain bags were effective in controlling the infestation, however, the insect mortality was not complete. Bags were perforated by a larger grain borer.
	Maize	Benin	150 days	Compared performance of hermetic bags and woven polypropylene bags for storage of maize infested with Prostephanus truncatus (Horn) and Sitophilus zeamaiswas (Motschulsky).	Moisture levels remained unchanged in hermetic bags. Growth of insects (Prostephanus truncatus and Sitophilus zeamaiswas) was significantly less in hermetic bags. There were 0.5%–6% losses at end of storage compared to 19.2%–27.1% losses in woven bags.
	Rice	Bangladesh	4 months	Compared performance of hermetic bags and traditional structures for storage of rice.	Moisture content of grains remained unchanged in hermetic bags. A total of 97% seed germination in hermetic bags vs. 95% in traditional storage. A total of 1% damaged grains in hermetic bags in contrary to 6% in traditional storage.

PICS bags	Mung bean, pigeon-pea	Kenya	6 months	Evaluated performance of hermetic bags for naturally and artificially infested (Callosobruchus maculatus (F.)) grains.	One hundred grain weight, infestation, and grain damage remain unchanged in hermetic bags. There was 60.3 to 76.9% damage in mung beans and 75.8%–95.7% grain damage for pigeonpeas stored in woven polypropylene.
	Maize	Benin, Burkina Faso and Ghana	6.5 months	Evaluated performance of hermetic bags for preserving maize quality during storage.	There was 95%–100% insect mortality in hermetic bags. PICS bags maintained the seed viability and germination.
	Maize	Kenya	6 months	Evaluated performance of hermetic bags for naturally and artificially infested (Prostephanus truncates) grains.	There was 0%–2% weight loss in PICS bags compared to 36.3%–47.7% weight loss in woven polypropylene bags. There was a 13%–20.1% reduction in germination for grains stored in PICS bags compared to a 54.1%–78.4% drop for grains stored in wove bags.
PICS bags	Bambara groundnut	Maradi, Niger	7 months	Evaluated performance of hermetic bags for preserving naturally infested Bambara groundnut quality during storage.	For highly infested grains, oxygen concentrations decreased significantly in hermetic bags contrary to unchanged in woven bags. Infestation level of C. maculatus in woven bags was 128 times higher than that of hermetic bags. There was a 34.8%–89.3% decrease in seed viability in woven bags, whereas, there was no change in grains stored in PICS bags. Abrasions were observed in inner HDPE bags.
	Pigeonpeas	India	8 months	Compared performance of hermetic bags vs. gunny bags for storage of pigeonpeas.	Germination of infested grains in gunny bags dropped to 44.5% compared to high germination (77%) for grains stored for 8 months in hermetic bags.
	Groundnuts	India	4 months	Evaluated performance of hermetic bags for preserving the quality of natural and artificial infested groundnuts.	There was a 0.8% decrease in seed weight for groundnut stored in hermetic bags compared to 7.2% in cloth bags. Only 1.4% weight loss for artificially infested groundnut stored in hermetic bags compared to 39.6% in cloth bag. There was 92.3% germination for artificially infested seeds stored in hermetic bags compared to only 10% in the case of cloth bags.

Figure illustrates the effectiveness of various storage options compared to traditional polypropylene bags in reducing the losses of maize grains after 90 days of storage . It can be clearly observed that losses in all new storage techniques were significantly lower than those in the traditional storage, with the minimum being in the case of metallic silos. A nationwide study on postharvest losses of rice in China reported 7%–13% grain losses at the rural household storage facilities, compared to only 0.2% losses at the national reserve level using scientific storage structures.

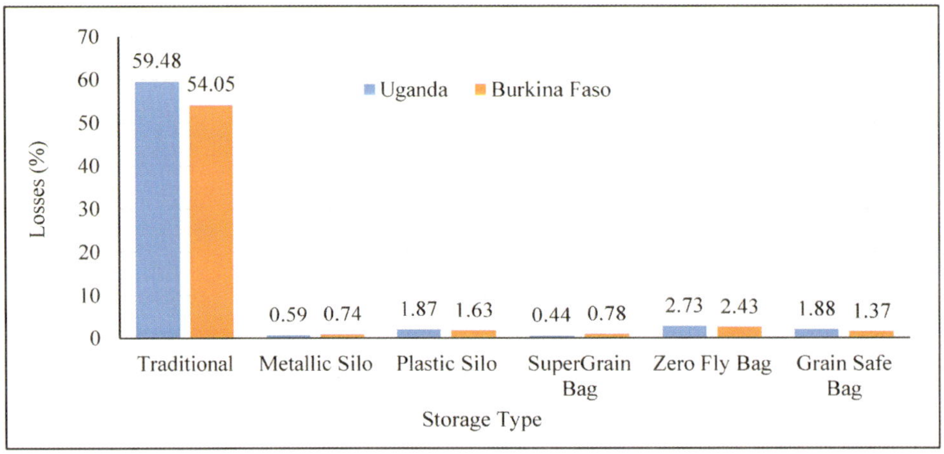

Losses in maize grain after 90 days of storage in various storage structures

Metal silos have been found to be effective in several other studies. However, their initial high cost is a major obstacle for their adoption by smallholders. Community level silos might be an economic alternative, as the cost per unit of grains decreases with increases in the size of silos. The maintenance cost is very low in the case of silos, which can compensate for the high initial cost to some extent. Kimenju and Hugo conducted an economic analysis of using advanced storage structures, and reported that the economic gain (extra income by avoiding losses) using a metal silo compared to polypropylene bags could be up to USD 100 per ton of grains after 12 months of storage. However, farmers have to spend an extra USD 171 (1.8-ton capacity) to USD 316 (0.36-ton capacity) as the initial cost of silos over polypropylene bags. One of the main challenges of using hermetic bags is that the grain to be stored should be thoroughly dried to avoid mold and rotting of grains. Although these bags prevent the damage from insects, they do not provide an effective barrier from rodents. Specifically, SuperGrain bags are widely used in Russia and Latin America for storage of coffee. They are also popularized in Afghanistan for wheat storage, Nepal for corn, and in Vietnam for rice conservation. These bags are being used successfully in several Latin American countries to store all major grain crops without application of pesticides. Hermetic storage structures developed by GrainPro Inc. are used by several commercial companies in India to store high value spices and Basmati rice.

Technology interventions and improved storage structures can significantly reduce store losses. However, it is important to understand that training of smallholders is equally as necessary as the technology dissemination. Along with making these technologies available at a reduced price, the government agencies and organizations have to ensure the development of facilities to provide information and training about the use and maintenance of these technologies in the local language, for successful adaptation and effective use of these technologies.

3

Postharvest Storage

This chapter closely examines the key processes and techniques of postharvest storage to provide an extensive understanding of agricultural postharvest technology. It includes various topics related to cold storage, frosting, freezing and chilling injury, temperature control with respect to different grains, fruits and vegetables and techniques of minimizing physical damage, besides others.

Postharvest Storage of Crops

If produce is to be stored, it is important to begin with a high quality product. The lot of produce must not contain damaged or diseased units, and containers must be well ventilated and strong enough to withstand stacking. In general proper storage practices include temperature control, relative humidity control, air circulation and maintenance of space between containers for adequate ventilation, and avoiding incompatible product mixes.

Commodities stored together should be capable of tolerating the same temperature, relative humidity and level of ethylene in the storage environment. High ethylene producers (such as ripe bananas, apples, cantaloupe) can stimulate physiological changes in ethylene sensitive commodities (such as lettuce, cucumbers, carrots, potatoes, sweet potatoes) leading to often undesirable color, flavor and texture changes.

Temperature management during storage can be aided by constructing square rather than rectangular buildings. Rectangular buildings have more wall area per square feet of storage space, so more heat is conducted across the walls, making them more expensive to cool. Temperature management can also be aided by shading buildings, painting storehouses white or silver to help reflect the sun's rays, or by using sprinkler systems on the roof of a building for evaporative cooling. The United Nations' Food and Agriculture Organization (FAO) recommends the use of ferrocement for the construction of storage structures in tropical regions, with thick walls to provide insulation. Facilities located at higher altitudes can be effective, since air temperature decreases as altitude increases. Increased altitude therefore can make evaporative cooling, night cooling and radiant cooling more feasible. Underground storage for citrus crops is common in Southern China, while in Northwest China, apples are stored in caves. This system was widely used in the U.S. during the early pert of this century.

Certain commodities, such as onions and garlic, store better in lower relative humidity environments. Curing these crops and allowing the external layers of tissue to dry out prior to handling and storage helps to protect them from further water loss.

The air composition in the storage environment can be manipulated by increasing or decreasing the rate of ventilation (introduction of fresh air) or by using gas absorbers such as potassium permanganate or activated charcoal. Large-scale controlled or modified atmosphere storage requires complex technology and management skills, however, some simple methods are available for handling small volumes of produce.

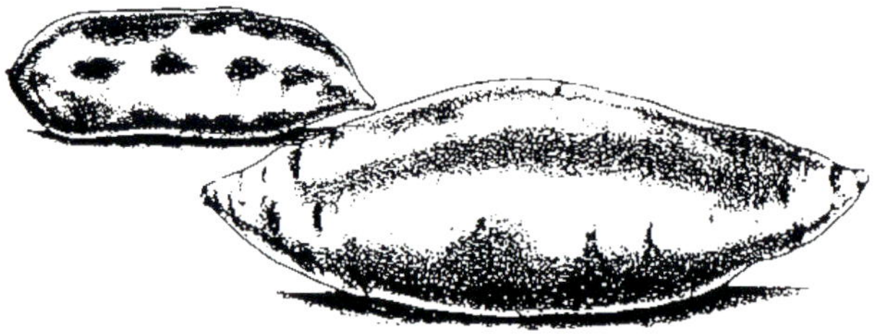

Recommended Storage Temperatures

Recommended temperature and relative humidity, and approximate transit and storage life for fruits and vegetable crops.

Product	Temperature		Relative Humidity (percent)	Approximate storage life
	°C	°F		
Amaranth	0-2	32-36	95-100	10-14 days
Anise	0-2	32-36	90-95	2-3 weeks
Apples	-1-4	30-40	90-95	1-12 months
Apricots	-0.5-0	31-32	90-95	1-3 weeks
Artichokes, globe	0	32	95-100	2-3 weeks
Asian pear	1	34	90-95	5-6 months
Asparagus	0-2	32-35	95-100	2-3 weeks
Atemoya	13	55	85-90	4-6 weeks
Avocados, Fuerte, Hass	7	45	85-90	2 weeks
Avocados, Lula, Booth-1	4	40	90-95	4-8 weeks
Avocados, Fuchs, Pollock	13	55	85-90	2 weeks
Babaco	7	45	85-90	1-3 weeks
Bananas, green	13-14	56-58	90-95	14 weeks
Barbados cherry	0	32	85-90	7-8 weeks
Bean sprouts	0	32	95-100	7-9 days
Beans, dry	4-10	40-50	40-50	6-10 months
Beans, green or snap	4-7	4045	95	7-10 days
Beans, lima, in pods	5-6	4143	95	5 days
Beets, bunched	0	32	98-100	10-14 days
Beets, topped	0	32	98-100	4-6 months
Belgian endive	2-3	36-38	95-98	24 weeks
Bitter melon	12-13	53-55	85-90	2-3 weeks
Black sapote	13-15	55-60	85-90	2-3 weeks
Blackberries	-0.5-0	31-32	90-95	2-3 days
Blood orange	4-7	4044	90-95	3-8 weeks
Blueberries	-0.5-0	31-32	90-95	2 weeks

Product	Temperature		Relative Humidity (percent)	Approximate storage life
	°C	°F		
Bok choy	0	32	95-100	3 weeks
Boniato	13-15	55-60	85-90	4-5 months
Breadfruit	13-15	55-60	85-90	2-6 weeks
Broccoli	0	32	95-100	10-14 days
Brussels sprouts	0	32	95-100	3-5 weeks
Cabbage, early	0	32	98-100	3-6 weeks
Cabbage, late	0	32	98-100	5-6 months
Cactus Leaves	24	3640	90-95	3 weeks
Cactus Pear	24	36-40	90-95	3 weeks
Caimito	3	38	90	3 weeks
Calabaza	10-13	50-55	50-70	2-3 months
Calamondin	9-10	48-50	90	2 weeks
Canistel	13-15	55-60	85-90	3 weeks
Cantaloups (3/4-slip)	2-5	36-41	95	15 days
Cantaloups (full-slip)	0-2	32-36	95	5-14 days
Carambola	9-10	48-50	85-90	3-4 weeks
Carrots, bunched	0	32	95-100	2 weeks
Carrots, mature	0	32	98-100	7-9 months
Carrots, immature	0	32	98-100	4-6 weeks
Cashew apple	0-2	32-36	85-90	5 weeks
Cauliflower	0	32	95-98	34 weeks
Celeriac	0	32	97-99	6-8 months
Celery	0	32	98-100	2-3 months
Chard	0	32	95-100	10-14 days
Chayote squash	7	45	85-90	4-6 weeks
Cherimoya	13	55	90-95	2-4 weeks
Cherries, sour	0	32	90-95	3-7 days
Cherries, sweet	-1 to -0.5	30-31	90-95	2-3 weeks
Chinese broccoli	0	32	95-100	10-14 days
Chinese cabbage	0	32	95-100	2-3 months
Chinese long bean	4-7	40-45	90-95	7-10 days
Clementine	4	40	90-95	24 weeks
Coconuts	0-1.5	32-35	80-85	1-2 months
Collards	0	32	95-100	10-14 days
Corn, sweet	0	32	95-98	5-8 days
Cranberries	2-4	36-40	90-95	24 months
Cucumbers	10-13	50-55	95	10-14 days
Currants	-0.5-0	31-32	90-95	1-4 weeks
Custard apples	5-7	41-45	85-90	4-6 weeks
Daikon	0-1	32-34	95-100	4 months
Dates	-18 or 0	0 or 32	75	6-12 months

Product	Temperature		Relative Humidity (percent)	Approximate storage life
	°C	°F		
Dewberries	-0.5-0	31-32	90-95	2-3 days
Durian	4-6	39-42	85-90	6-8 weeks
Eggplants	12	54	90-95	1 week
Elderberries	-0.5-0	31-32	90-95	1-2 weeks
Endive and escarole	0	32	95-100	2-3 weeks
Feijoa	5-10	41-50	90	2-3 weeks
Figs fresh	-0.5-0	31-32	85-90	7-10 days
Garlic	0	32	65-70	6-7 months
Ginger root	13	55	65	6 months
Gooseberries	-0.5-0	31-32	90-95	34 weeks
Granadilla	10	50	85-90	3-4 weeks
Grapefruit, Calif. & Ariz.	14-15	58-60	85-90	6-8 weeks
Grapefruit, Fla. & Texas	10-15	50-60	85-90	6-8 weeks
Grapes, Vinifera	-1 to -0.5	30-31	90-95	1-6 months
Grapes, American	-0.5-0	31-32	85	2-8 weeks
Greens, leafy	0	32	95-100	10-14 days
Guavas	5-10	41-50	90	2-3 weeks
Haricot vert	4-7	4045	95	7-10 days
Horseradish	-1-0	30-32	98-100	10-12 months
Jaboticaba	13-15	55-60	90-95	2-3 days
Jackfruit	13	55	85-90	2-6 weeks
Jaffa orange	8-10	46-50	85-90	8-12 weeks
Japanese eggplant	8-12	46-54	90-95	1 week
Jerusalem Artichoke	-0.5-0	31-32	90-95	+5 months
Jicama	13-18	55-65	65-70	1-2 months
Kale	0	32	95-100	2-3 weeks
Kiwano	10-15	50-60	90	6 months
Kiwifruit	0	32	90-95	3-5 months
Kohlrabi	0	32	98-100	2-3 months
Kumquats	4	40	90-95	2-4 weeks
Langsat	11-14	52-58	85-90	2 weeks
Leeks	0	32	95-100	2-3 months
Lemons	10-13	50-55	85-90	1-6 months
Lettuce	0	32	98-100	2-3 weeks
Limes	9-10	48-50	85-90	6-8 weeks
Lo bok	0-1.5	32-35	95-100	24 months
Loganberries	-0.5-0	31-32	90-95	2-3 days
Longan	1.5	35	90-95	3-5 weeks
Loquats	0	32	90	3 weeks
Lychees	1.5	35	90-95	3-5 weeks
Malanga	7	45	70-80	3 months

Product		Temperature		Relative Humidity (percent)	Approximate storage life
		°C	°F		
Mamey		13-15	55-60	90-95	2-6 weeks
Mangoes		13	55	85-90	2-3 weeks
Mangosteen		13	55	85-90	2-4 weeks
Melons:					
	Casaba	10	50	90-95	3 weeks
	Crenshaw	7	45	90-95	2 weeks
	Honeydew	7	45	90-95	3 weeks
	Persian	7	45	90-95	2 weeks
Mushrooms		0	32	95	34 days
Nectarines		-0.5-0	31-32	90-95	2-4 weeks
Okra		7-10	45-50	90-95	7-10 days
Olives, fresh		5-10	41-50	85-90	+6 weeks
Onions, green		0	32	95-100	34 weeks
Onions, dry		0	32	65-70	1-8 months
Onion sets		0	32	65-70	6-8 months
Oranges, Calif. & Ariz.		3-9	3848	85-90	3-8 weeks
Oranges, Fla. & Texas		0-1	32-34	85-90	8-12 weeks
Papayas		7-13	45-55	85-90	1-3 weeks
Passionfruit		7-10	45-50	85-90	3-5 weeks
Parsley		0	32	95-100	2-2.5 months
Parsnips		0	32	95-100	+6 months
Peaches		-0.5-0	31-32	90-95	2-4 weeks
Pears		-1.5 to -0.5	29-31	90-95	2-7 months
Peas, green		0	32	95-98	1-2 weeks
Peas, southern		+5	4041	95	6-8 days
Pepino		4	40	85-90	1 month
Peppers, Chili (dry)		0-10	32-50	60-70	6 months
Peppers, sweet		7-13	45-55	90-95	2-3 weeks
Persimmons, Japanese		-1	30	90	34 months
Pineapples		7-13	45-55	85-90	24 weeks
Plantain		13-14	55-58	90-95	1-5 weeks
Plums and prunes		-0.5-0	31-32	90-95	2-5 weeks
Pomegranates		5	41	90-95	2-3 months
Potatoes, early crop		10-16	50-60	90-95	10-14 days
Potatoes, late crop		4.5-13	40-55	90-95	5-10 months
Pummelo		7-9	4548	85-90	12 weeks
Pumpkins		10-13	50-55	50-70	2-3 months
Quinces		-0.5-0	31-32	90	2-3 months
Raddichio		0-1	32-34	95-100	2-3 weeks
Radishes, spring		0	32	95-100	34 weeks
Radishes, winter		0	32	95-100	24 months

Product	Temperature		Relative Humidity (percent)	Approximate storage life
	°C	°F		
Rambutan	12	54	90-95	1-3 weeks
Raspberries	-0.5-0	31-32	90-95	2-3 days
Rhubarb	0	32	95-100	24 weeks
Rutabagas	0	32	98-100	+6 months
Salsify	0	32	95-98	2-4 months
Santol	7-9	45-48	85-90	3 weeks
Sapodilla	16-20	60-68	85-90	2-3 weeks
Scorzonera	0-1	32-34	95-98	6 months
Seedless cucumbers	10-13	50-55	85-90	10-14 days
Snow peas	0-1	32-34	90-95	1-2 weeks
Soursop	13	55	85-90	1-2 weeks
Spinach	0	32	95-100	10-14 days
Squashes, summer	5-10	41-50	95	1-2 weeks
Squashes, winter	10	50	50-70	2-3 months
Strawberries	0	32	90-95	5-7 days
Sugar apples	7	45	85-90	4 weeks
Sweetpotatoes	13-15	55-60	85-90	4-7 months
Tamarillos	3-4	37-40	85-95	10 weeks
Tamarinds	7	45	90-95	3-4 weeks
Tangerines, mandarins, and related citrus fruits	4	40	90-95	24 weeks
Taro root	7-10	45-50	85-90	4-5 months
Tomatillos	13-15	55-60	85-90	3 weeks
Tomatoes, mature-green	18-22	65-72	90-95	1-3 weeks
Tomatoes, firm-ripe	13-15	55-60	90-95	4-7 days
Turnips	0	32	95	4-5 months
Turnip greens	0	32	95-100	10-14 days
Ugli fruit	4	40	90-95	2-3 weeks
Waterchestnuts	0-2	32-36	98-100	1-2 months
Watercress	0	32	95-100	2-3 weeks
Watermelons	10-15	50-60	90	2-3 weeks
White sapote	19-21	67-70	85-90	2-3 weeks
White asparagus	0-2	32-36	95-100	2-3 weeks
Winged bean	10	50	90	4 weeks
Yams	16	61	70-80	6-7 months
Yucca root	0-5	32-41	85-90	1-2 months

Compatibility Groups for Storage of Fruits, Vegetables and Floral Crops

Group 1: Fruits and vegetables, 0 to 2° C (32 to 36° F), 90-95% relative humidity. Many products in this group produce ethylene.

apples	grapes (without	parsnips
apricots	sulfur dioxide)	peaches
Asian pears	horseradish	pears
Barbados cherry	kohlrabi	persimmons
beets, topped	leeks	plums
berries (except	longan	pomegranates
cranberries)	loquat	prunes
cashew apple	lychee	quinces
cherries	mushrooms	radishes
coconuts	nectarines	rutabagas
figs (not with apples)	oranges* (Florida and Texas)	turnips

Group 2: Fruits and vegetables, 0 to 2° C (32 to 36° F), 95-100% relative humidity. Many products in this group are sensitive to ethylene.

Amaranth*	corn, sweet*	parsley*
anise	daikon*	parsnips*
artichokes*	endive*	peas*
asparagus	escarole*	pomegranate
bean sprouts	grapes (without sulfur dioxide)	raddichio
beets*		radishes*
Belgian endive	horseradish	rhubarb
berries (except cranberries)	Jerusalem artichoke	rutabagas*
	kiwifruit	salsify
bok choy	kohlrabi*	scorzonera
broccoli*	leafy greens	snow peas
brussels sprouts*	leeks' (not with figs or grapes)	spinach*
cabbage*		turnips*
carrots*	lettuce	waterchestnut
cauliflower	lo bok	watercress*
celeriac*	mushrooms	
celery*	onions, green* (not with figs, grapes, mushrooms, rhubarb, or corn)	
cherries		

Group 3: Fruits and vegetables, 0 to 2° C (32 to 36° F), 65-75% relative humidity. Moisture will damage these products.

Garlic
onions, dry

Group 4: Fruits and vegetables, 4.5° C (40° F), 90-95% relative humidity.

cactus leaves	lemons*	tamarillo
cactus pears	lychees	tangelos*
caimito	kumquat	tangerines*
cantaloupes**	mandarin*	ugli fruit*
clementine	oranges (Calif. and Arizona)	yucca root
cranberries	pepino	

Group 5: Fruits and vegetables, 10° C (50° F), 85-90% relative humidity. Many of these products are sensitive to ethylene. These products also are sensitive to chilling injury.

beans	kiwano	pummelo
calamondin	malanga	squash, summer
chayote	okra	(sot shell)
cucumber	olive	tamarind
eggplant	peppers	taro root
haricot vert	potatoes, storage	

Group 6: Fruits and vegetables, 13 to 15° C (55 to 60° F), 85-90% relative humidity. Many of these products produce ethylene. These products also are sensitive to chilling injury.

atemoya	granadilla	papayas
avocados	grapefruit	passionfruit
babaco	guava	pineapple
bananas	jaboticaba	plantain
bitter melon	jackfruit	potatoes, new
black sapote	langsat	pumpkin
boniato	lemons*	rambutan
breadfruit	limes*	santol
canister	mamey	soursop
carambola	mangoes	sugar apple
cherimoya	mangosteen	squash, winter
coconuts	melons (except cantaloupes)	(hard shell)
feijoa		tomatillos
ginger root		tomatoes, ripe

Group 7: Fruits and vegetables, 18 to 21° C (65 to 70° F), 85-90% relative humidity.

jicama	sweetpotatoes*	watermelon*
pears	tomatoes,	white sapote
(for ripening)	mature green	yams*

Group 8: Flowers and florist greens, 0 to 2° C (32 to 36° F), 90-95% relative humidity.

allium	freesia	peony, tight
aster, China	gardenia	buds
bouvardia	hyacinth	ranunculus
carnation	iris, bulbous	rose
chrysanthemum	lily	squill
crocus	lily-of-the-valley	sweet pea
cymbidium orchid	narcissus	tulip
adiantum (maidenhair)	ground pine	rhododendren
cedar	flex (holly)	salal (lemon leaf)
dagger and wood	juniper	
ferns	mistletoe	vaccinium
galax	mountain-laurel	(huckleberry)
woodwardia fern		

Group 9: Flowers and florist greens, 4.5° C (40° F), 90-95% relative humidity.

acacia	delphinium	orchid,
alstromeria	feverfew	cymbidium
anemone	forget-me-not	ornithogalum
aster, China	foxglove	poppy
buddleia	gaillardia	phlox
calendula	gerbera	primrose
calla	gladiolus	protect
candytuft	gloriosa	ranunculus
clarkia	gypsophilla	snapdragon
columbine	heather	snowdrop
coreopsis	laceflower	statice
cornflower	lilac, forced	stephanotis
cosmos	lupine	stevia
dahlia	marigolds	stock
daisies	mignonette	strawflower
violet	zinnia	
adiantum (maidenhair)	eucalyptus	myrtus (myrtle)
asparagus (plumosa, sprenger)	hedera	philodendren
	flex (holly)	pittosporum
buxus (boxwood)	leatherleaf (baker fern)	pothos
camellia		scotch-broomern
croton	leucothoe, drooping	smilax, southern
dracaena	magnolia	woodwardia fern

Group 10: Flowers and florist greens, 7 to 10° C (45 to 50° F), 90-95% relative humidity.

anemone	eucharis	orchid, cattleya
bird-of-paradise	gloriosa	sweet william
camellia	godetia	
chamaedora	cordyline (ti)	palm
	podocarpus	

Group 11: Flowers and florist greens, 13 to 15° C (55 to 60° F), 90-95% relative humidity.

anthurium	heliconia	poinsetta
ginger	orchid, vanda	
diffenbachia	stag horn fern	

Storage Practices

Inspecting stored produce and cleaning storage structures on a regular basis will help reduce losses by minimizing the buildup of pests and discouraging the spread of diseases.

Inspect produce and clean the storage structure:

Clean and maintain the storage structure:

Storage facilities should be protected from rodents by keeping the immediate area clean, free from trash and weeds. Rat guards can be made from simple materials such as old tin cans or pieces of sheet metal fashioned to fit the extended legs of storage structures. If desired, more elaborate technologies can be used. Concrete floors will help prevent rodent entry, as will screens on windows, vents and drains.

Remove trash and weeds:

Cement floors:

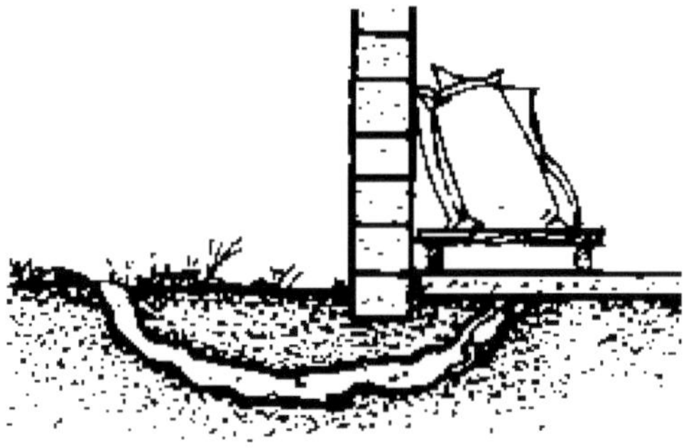

When inspecting stored produce, any spoiled or infected produce should be removed and destroyed. In some cases, produce may still be fit for consumption if used immediately perhaps as animal feed. Reusable containers and sacks should be disinfected in chlorinated or boiling water before reuse.

Disinfect used sacks:

Placing materials on the floor beneath sacks or cartons of produce prevents dampness from reaching produce suited to dry conditions in storage. This helps to reduce the chance of fungal infection, while also improving ventilation and sanitation in the storeroom. Some examples of useful materials follow.

Waterproof sheets:

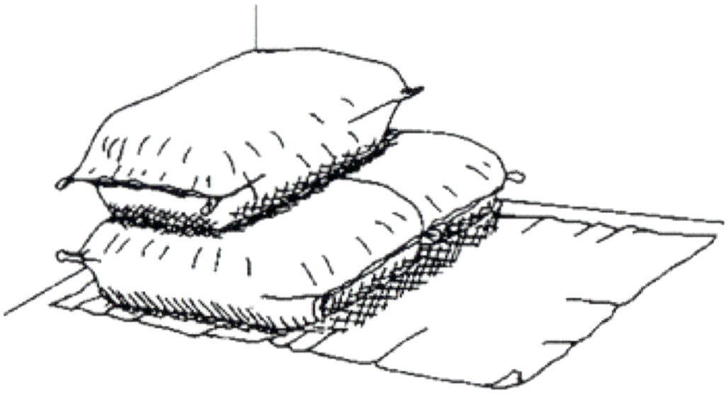

Poles:

Wooden pallets:

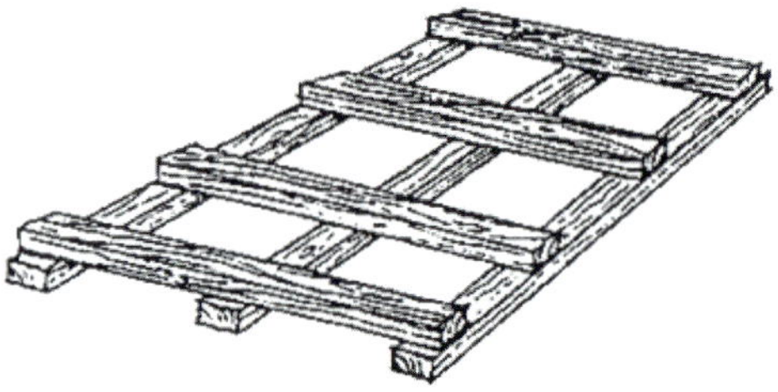

Storage Structures

A yam barn is a traditional structure used in West Africa to store yams after curing. Fast-growing, live trees are used to create a rectangular structure, and form the framework of the barn as well as provide shade.

Outside view of barn with 'live' shade,

Trunk of fast growing tree-planted in situ.

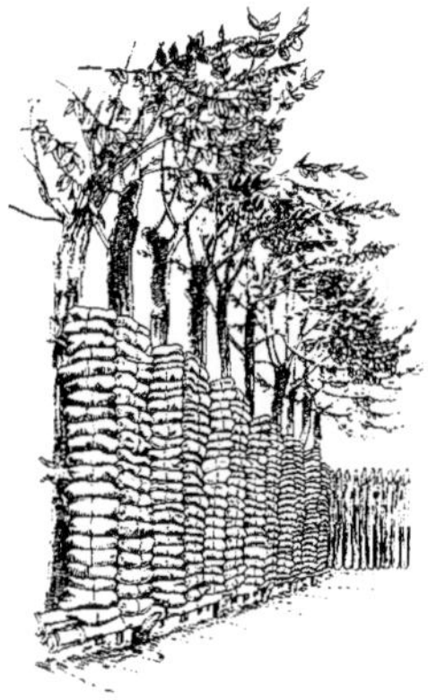

Inside of barn showing tying of yams,

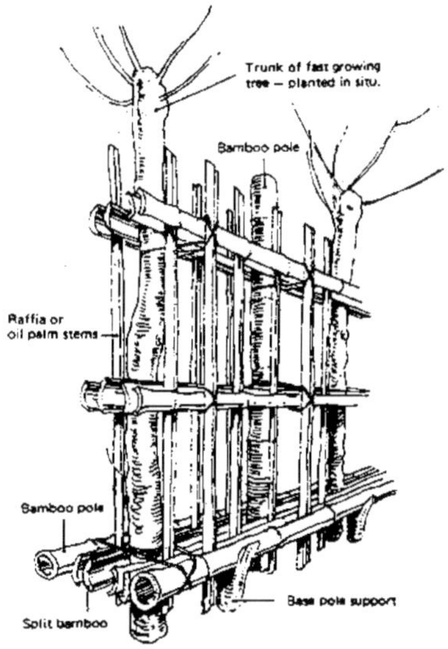

Storage facilities require adequate ventilation in order to help extend shelf life and maintain produce quality. The following are three types of fans found in common use.

Centrifugal:

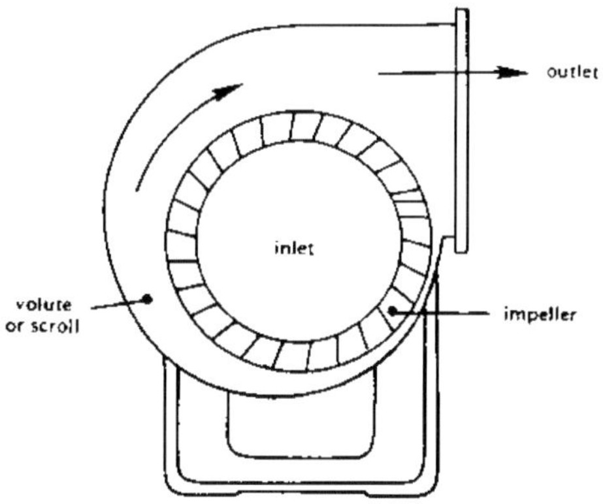

Axial flow:

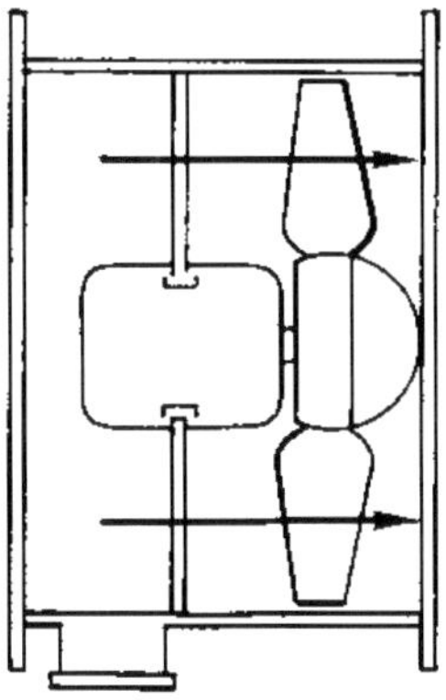

Propeller/expeller:

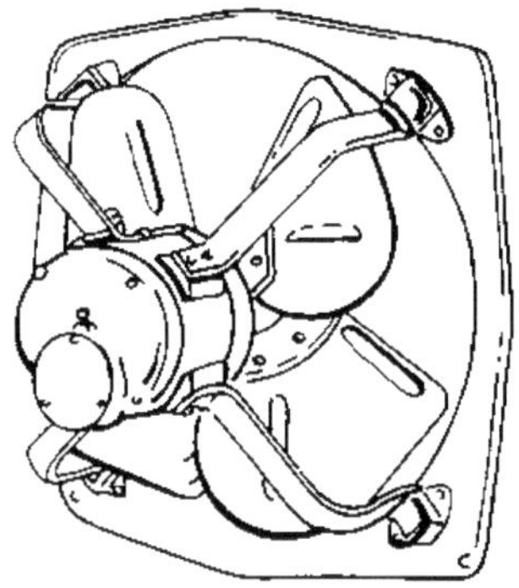

Ventilation in storage structures is improved if air inlets are located at the bottom of the store, while air outlets are at the top. A simple, light-proof exhaust vent is a pressure-relief flap.

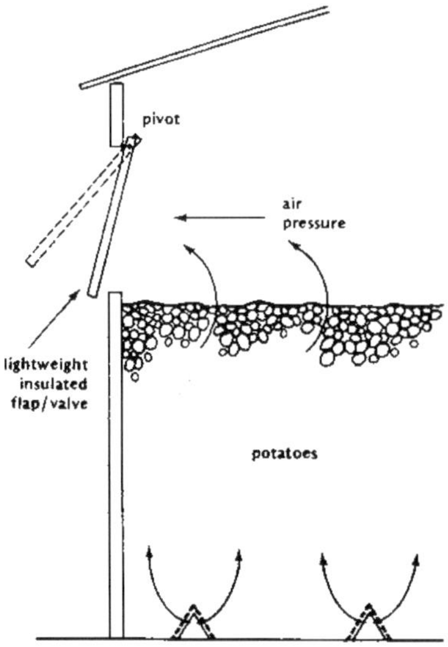

Any type of building or facility used for storage of Crops should be insulated for maximum effectiveness. A well insulated refrigerated building will require less electricity to keep produce cool. If the structure is to be cooled by evaporative or night air ventilation, a well insulated building will hold the cooled air longer.

Insulation R-values are listed below for some common building materials. R refers to resistance, and the higher the R-value, the higher the material's resistance to heat conduction and the better the insulating property of the material.

R – Value

Material	1 inch thick
Batt and Blanket Insulation	
Glass wool, mineral wool, or fiberglass	3.50
Fill-Type Insulation	
Cellulose	3.50
Glass or mineral wool	2.50-3.00
Vermiculite	2.20
Wood shavings or sawdust	2.22
Rigid Insulation	
Plain expanded extruded polystyrene	5.00
Expanded rubber	4.55
Expanded polystyrene molded beads	3.57
Aged expanded polyurethane	6.25
Glass fiber	4.00
Polyisocyranuate	8.00
Wood or cane fiber board	2.50
Foamed-in-Place Insulation	
Sprayed expanded urethane	6.25

Building Materials	Full thickness of material
Solid concrete	0.08
8-inch concrete block, open core	1.11
8-inch lightweight concrete block open core	2.00
8-inch concrete block with vermiculite in core	5.03
Lumber, fir or pine	1.25
Metal siding	<0.01
3/8-inch plywood	1.25 - 0.47
1/2-inch plywood	1.25 - 0.62
Masonite particleboard	1.06
25/32-inch insulated sheathing	2.06
1/2-inch Sheetrock	0.45
1/2-inch wood lapsiding	0.81

An evaporative cooler located in the peak of a storage structure can cool an entire room of stored produce such as sweetpotatoes or other chilling sensitive crops. The vents for outside air should be located at the base of the building so that cool air is circulated throughout the room before it can exit.

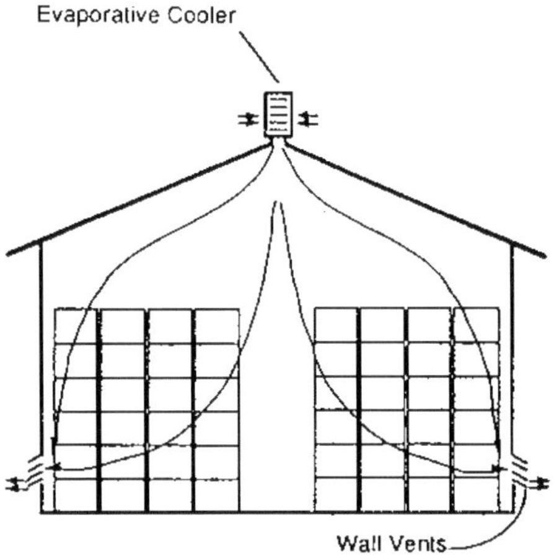

Illustrated below is a cross-sectional view of a storehouse for fruits. This system was officially approved as the standard model for farm-level storehouses by the Ministry of Construction (Korea) in 1983. Note that air inlets are at the base of the building, and the floor is perforated, allowing free movement of air. The entire building is set below ground level taking advantage of the cooling properties of soil.

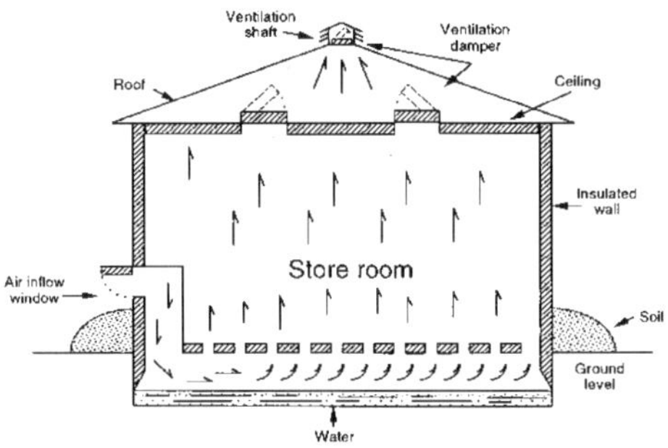

Commercially constructed cold rooms can be quite expensive, but fortunately the small-scale operator has many choices. Cold rooms can be self-constructed, purchased as prefabricated units (new or used), or made from refrigerated transportation equipment such as railway cars, highway vans or marine containers. Illustrated below is the basic plan for a self-built cold room. For more detailed information about determining the cold room size best suited to your operation, evaluating choices when purchasing or building a cold room, refer to the source below,

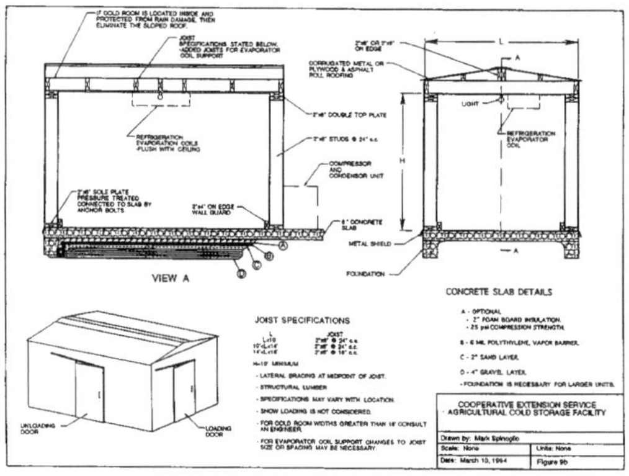

For storage facilities that are refrigerated, using outside air for ventilation is wasteful of energy.

For these systems, a simple recirculation system can be designed by adding a fan below floor level and providing a free space at one end of the storeroom for cool air to return to the inlet vents.

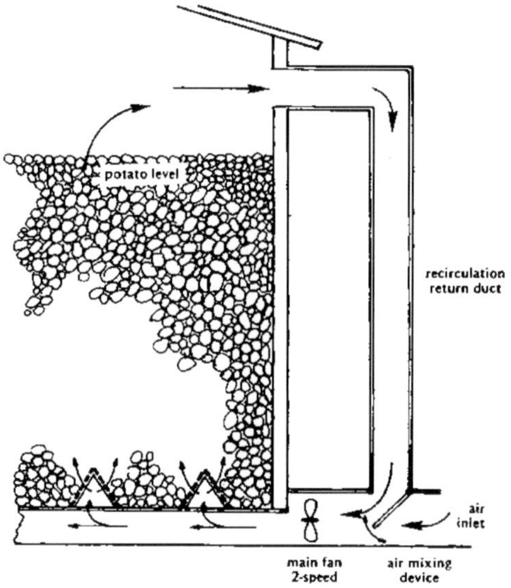

The proper arrangement of floor vents for air circulation will improve ventilation in the storage house. Lateral ducts should be 2 meters apart when measured from center to center. Air How velocities from the main duct should be 10 to 13 meters/second.

Longitudinal main duct:

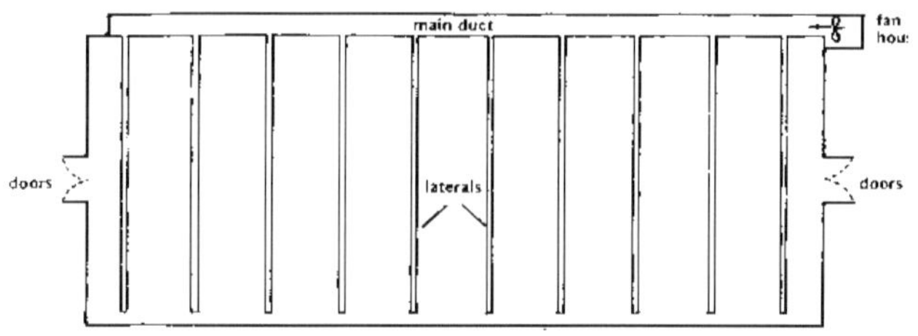

Central main duct:

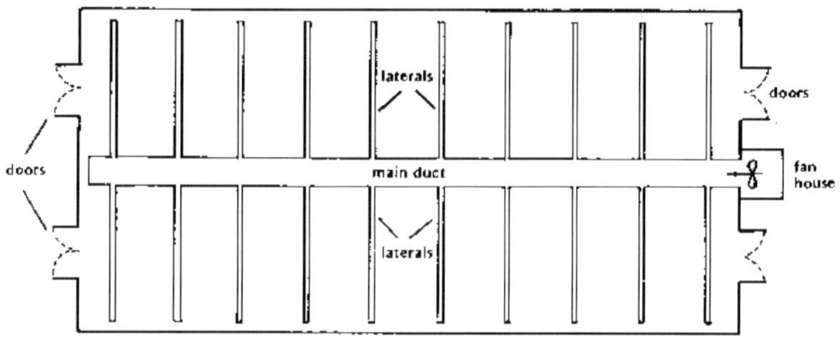

Lateral ducts can be constructed of a variety of materials. Portable vents can be made from wooden slats, in a triangular, square or rectangular design. A round tube of plastic or clay can be used if holes can be drilled without causing structural damage, or permanent ducts can be constructed below ground, using concrete blocks.

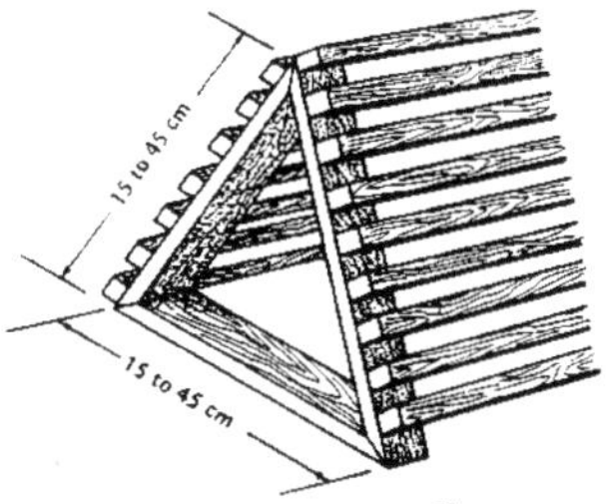

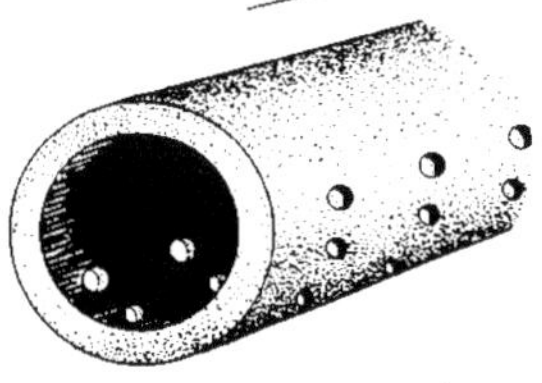

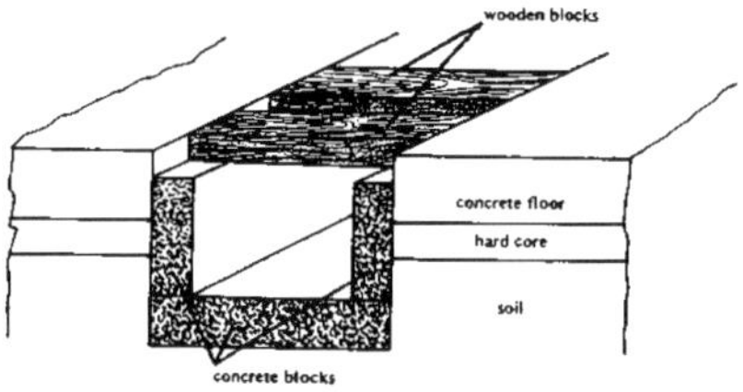

In cooler regions, suitable storage temperatures can be maintained by bringing outside air into the storage facility. Typical fan installations for a pressurized ventilation system are illustrated below. Overhead distribution of air simplifies the storehouse design. A indoor recirculation inlet can be added if refrigeration is in use. Ducts can be constructed of wood, plastic tubing or any suitable materials.

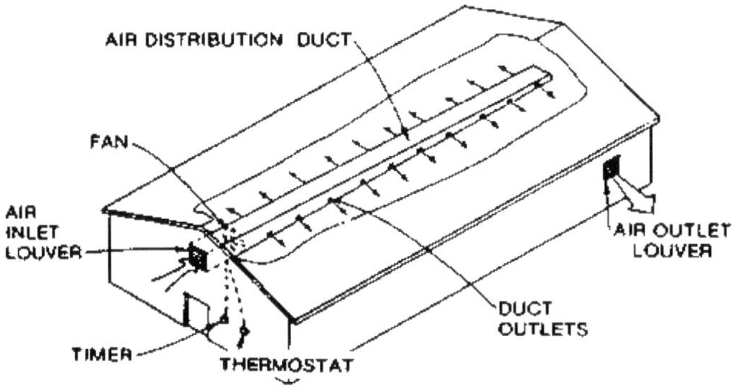

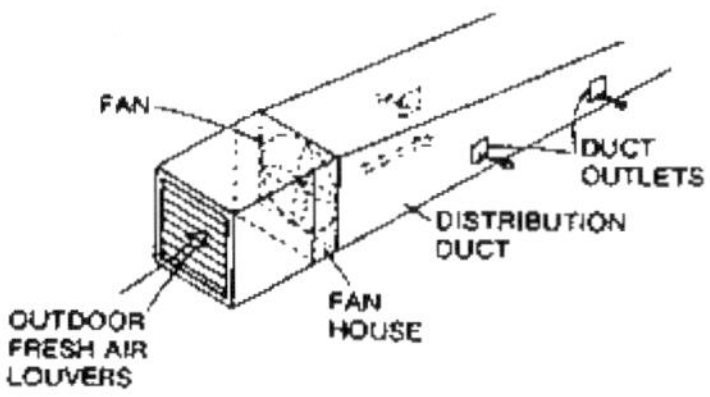

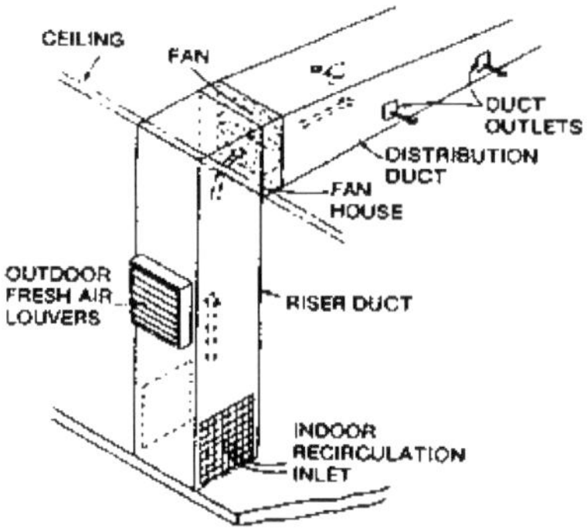

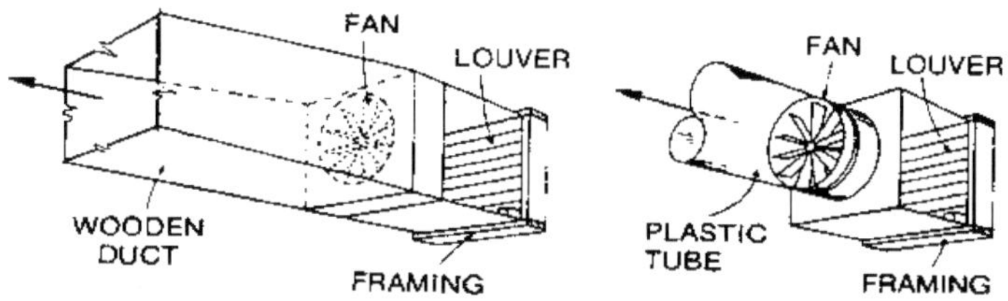

Storage structures can be cooled by ventilating at night when outside air is cool. For best results, air vents should be located at the base of the storage structure. An exhaust fan located at the top of the structure pulls the cool air through the storeroom. Vents should be closed at sunrise, and remain closed during the heat of the day.

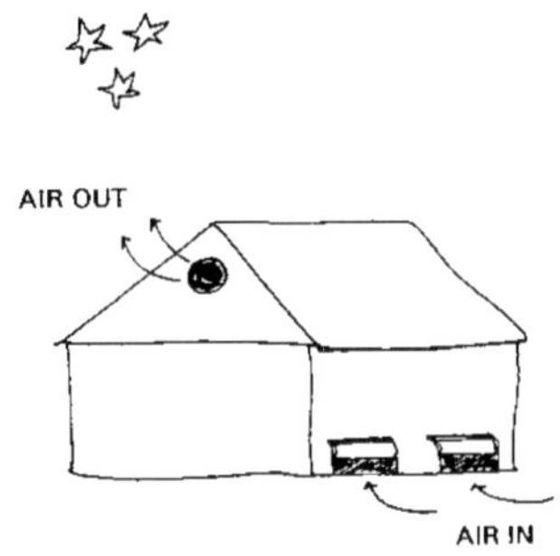

Overhanging roof extensions on storage structures are very helpful in shading the walls and ventilation openings from the sun's rays, and in providing protection from rain. An overhang of at least 1 meter (3 feet) is recommended.

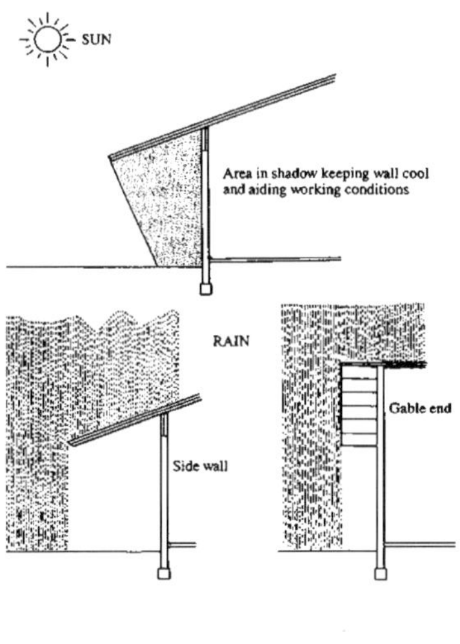

Where electricity is not available, wind-powered turbines can help keep storerooms cool by pulling air up through the building. Vents at the floor level are especially useful for cooling via night air ventilation.

The turbine illustrated below can be constructed of sheet metal that is twisted to catch the wind, and attached to a central pole that acts as the axis of rotation. Warm air in the storage room rises, causing the turbine to rotate, expelling the air and initiating an upward flow of warm air. The turbine should be placed on the peak of the roof of a storage structure

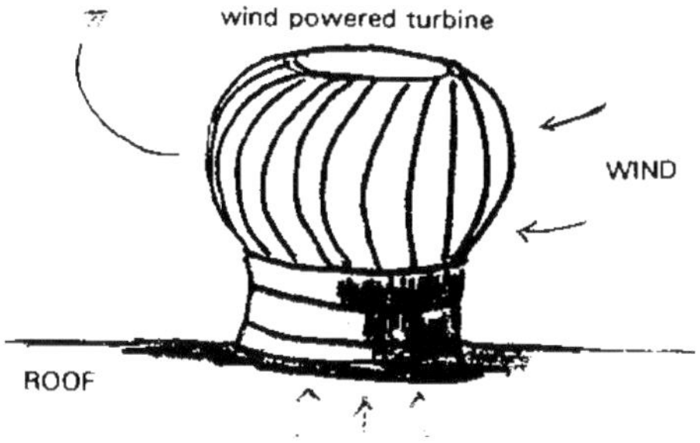

Protected surface storage is a simple method for storing small quantities of produce. The examples illustrated below are especially good for storage when night temperatures are lower than that desired for proper storage. Insulating materials such as straw can be used and protective covers can be constructed from wooden planks, plastic sheeting or layers of compacted soil.

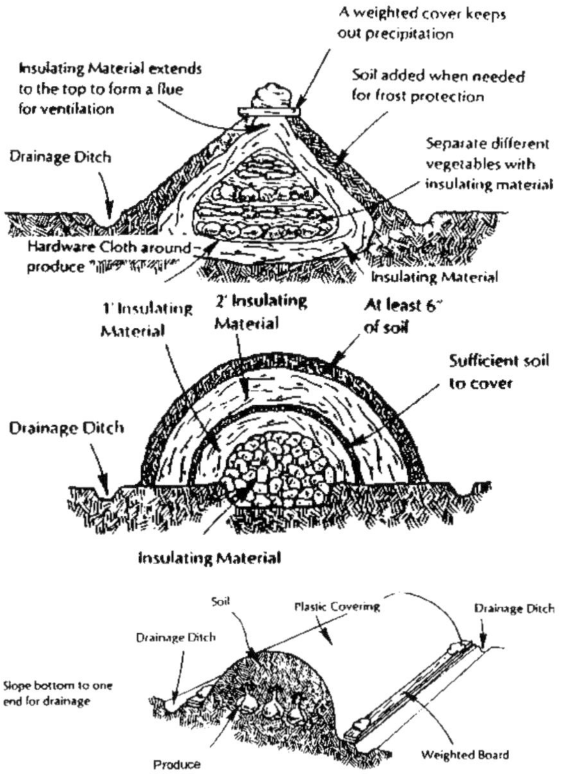

One of the simplest methods for storing small quantities of produce is to use any available container, and create a cool environment for storage by burying the container using insulating materials and soil. The example provided here employs a wooden barrel and straw for insulation.

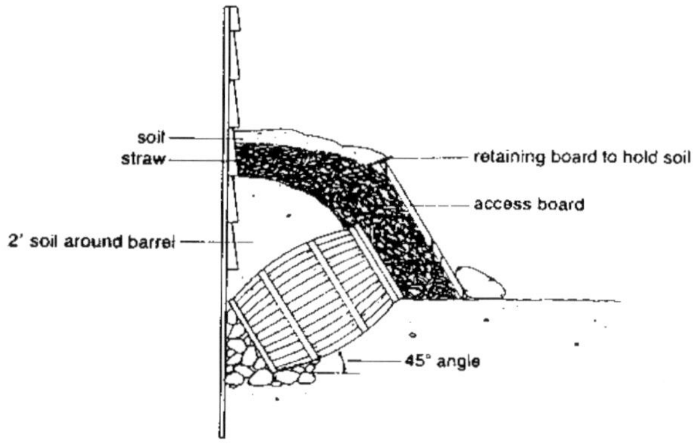

A root cellar can be constructed by digging out a pit to a depth of about 2 meters (7 to 8 feet) and framing the sides with wooden planks. The example illustrated here is about 3 by 4 meters (12 by 15 feet) in size, with a 35 cm square (one foot square) wooden chute as a roof vent.

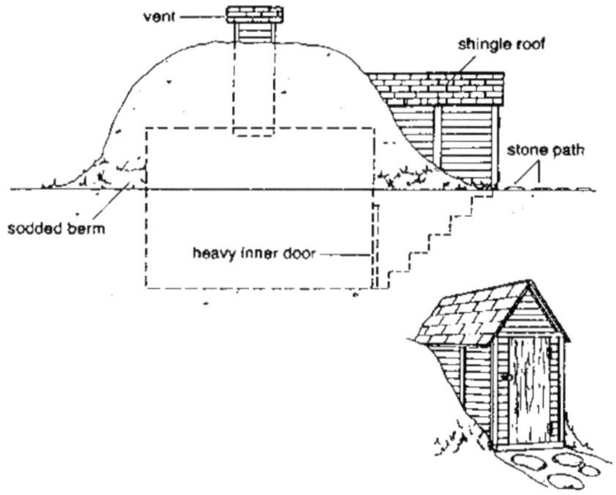

An outdoor storage bin can serve as a place to keep small quantities of potatoes in a region with a cool but not freezing climate. A wooden rack provides an air space for ventilation and straw provides insulation. The best location for such a structure would be in a shady spot.

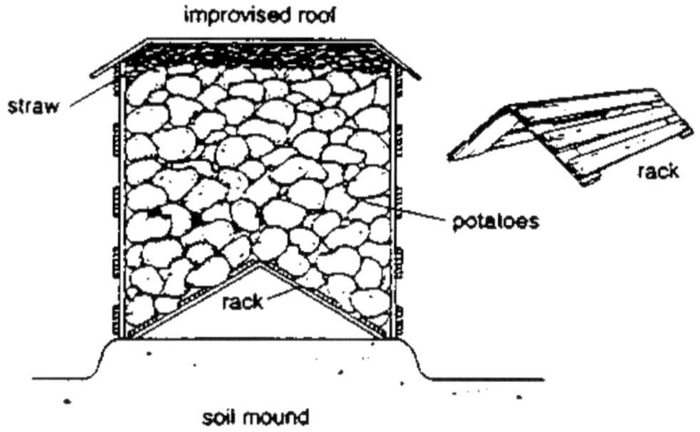

A root box, lined with hardware cloth and straw, buried to the top edge in soil will keep potatoes cool wile providing protection from freezing. The wooden lid can be lifted for easy access to produce, and straw bales on top provide more insulation.

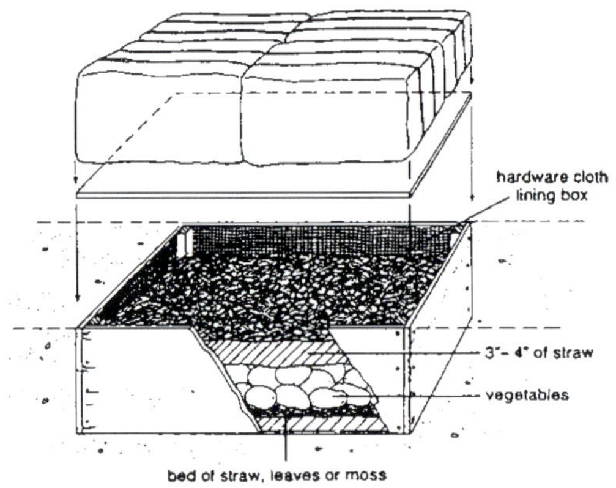

Dried and Bulb Crops

Onions, garlic and dried produce are best suited to low humidity in storage. Onions and garlic will sprout if stored at intermediate temperatures. Pungent types of onions will store longer than mild onions, which are rarely stored for more than one month. The following table lists the storage conditions recommended for these crops:

	Temperature		RH	Potential storage duration
	C	F	%	
Onions	0-5	32-41	65-70	6-8 months
	28-30	82-86	65-70	1 month
Garlic	0	32	70	6-7 months
	28-30	82-86	70	1 month
Dried fruits and vegetables	<10	<50	55-60	6-12 months

For bulk storage of onions or garlic, ventilation systems should be designed to provide air into the store from the bottom of the room at a rate of 2 cubic feet per minute per cubic feet of produce. If produce is in cartons or bins, stacks must allow free movement of air. Rows of containers should be stacked parallel to the direction of the flow of air and be spaced six to seven inches apart. An adequate air supply must be provided at the bottom of each row and containers must be properly vented.

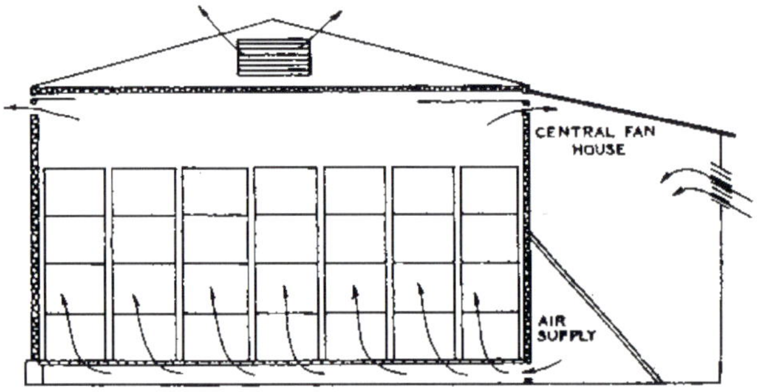

Root and Tuber Crops

The recommended storage conditions for root and tuber crops are listed in the following table. Potatoes for processing are best kept at intermediate temperatures to limit the production of sugars which darken when heated during processing. Potatoes meant for consumption must also be stored in the dark, since the tubers will produce chlorophyll (turning green) and develop the toxic alkaloid solanine if kept in the light. Potatoes stored for use as "seed" are best stored in diffuse light. The chlorophyll and solanine that accumulate will aid to protect the seed potatoes from insect pests and decay organisms.

Tropical root and tuber crops must be stored at temperatures that will protect the crops from chilling, since chilling injury can cause internal browning, surface pitting and increased susceptibility to decay.

	Temperature		RH%	Potential storage duration
	°C	°F		
Potatoes				
Fresh market	4-7	39-45	95-98	10 months
Processing	8-12	47-54	95-98	10 months
Seed potatoes	0-2	32-36	95-98	10 months
Cassava	5-8	41-47	80-90	2-4 months
	0-5	32-41	85-95	6 months
Sweet potato	12-14	54-58	85-90	6 months
Yam	13-15	55-59	Near 100	6 months
	27-30	80-86	60-70	3-5 months
Ginger	12-14	54-58	65-75	6 months
Jicama	12-15	54-59	65-75	3 months
Taro	13-15	55-59	85-90	4 months

Potatoes

When storing potatoes, a field storage clamp is a low cost technology that can be designed using locally available materials for ventilation and insulation. The example illustrated here employs a wooden ventilator box and straw for insulation. The entire pile of potatoes and straw is covered with a layer of soil, which should not be highly compacted. In very cold regions, a second layer of straw and soil can be added. In hot regions, less soil is needed, but more ventilation can be added by constructing chimney type air outlets at the top of the clamp.

Simple storage houses for potatoes can be constructed for small quantities of produce. The examples provided here can store 1 to 2 metric tons, and are used on farms and in mountain villages. The first is made from unfinished wooden planks painted white to reduce heat accumulation from the sun and covered with a large thatched roof for protection from sun and rain. It has a large door on one side for loading and unloading.

The second storage house is constructed from lath and plaster and mud bricks in a cylindrical form. It has two doors, one on top for loading, the other at the bottom for easy removal of potatoes for sale or consumption. White-wash helps reduce heat accumulation and a thatch roof protects the potatoes from rain and sun.

For large quantities of potatoes, a self-supporting A-frame storehouse can be constructed. A pit is dug about 10 feet deep and wooden air ducts are placed along the earthen floor. The roof of the building is constructed of wood, then covered with straw and soil.

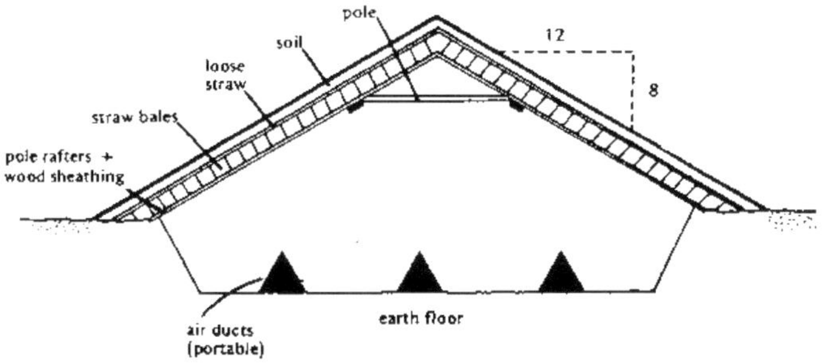

Ducts for ventilation of bulk storage rooms can be laid out vertically as well as horizontally. The storeroom for potatoes shown below provides for plenty of ventilation using simple materials. The room can be of any size or shape since air ducts can be positioned to extend evenly throughout.

When loading potatoes into bulk storage, even distribution of the produce is important for proper ventilation. Uneven loads will inhibit air movement and result in storage losses due to inadequate ventilation.

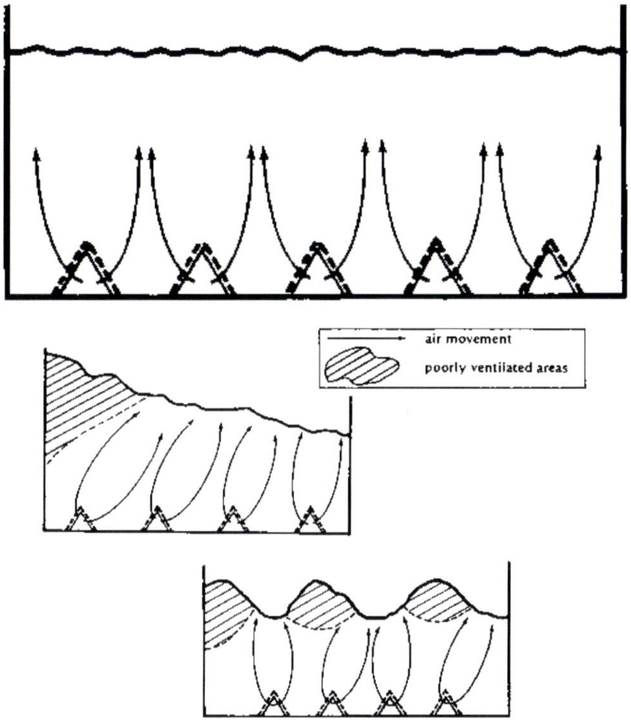

Controlled Atmosphere (C.A.) Storage

Controlled or modified atmosphere storage should be used as a supplement to, and not as a substitute for, proper temperature and relative humidity management. Some simple methods for modifying the composition of air in the storage environment are listed below. Air coming into the storeroom or being recirculated within the room must pass through a monitoring and control system.

- Oxygen gas control:

To Decrease:

  - Purging with nitrogen
  - From liquid nitrogen through an evaporator
  - From a membrane system nitrogen generator
  - From a molecular sieve system nitrogen generator

- Carbon dioxide control:

To INCREASE:

  - Dry ice
  - Pressurized gas cylinder

To DECREASE:

- Molecular sieve scrubber
- Activated charcoal scrubber
- Sodium hydroxide scrubber
- Hydrated lime (use 0.6 kg of hydrated lime to treat the air used to ventilate 100 kg of fruit Air can be directed to pass through a box, located inside or outside the C.A. storeroom).

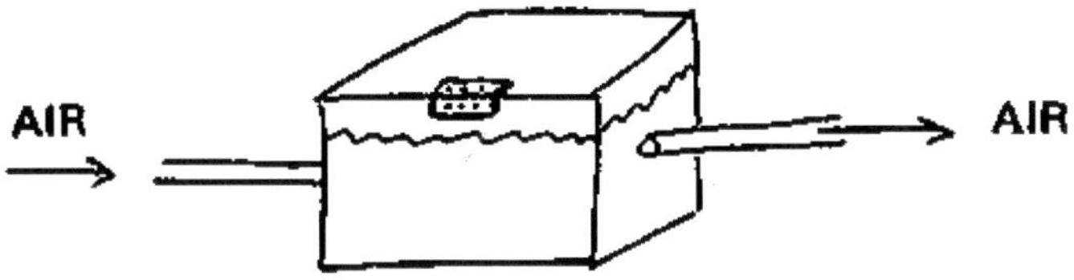

- Ethylene control:

To DECREASE:

- Potassium permanganate
- Activated charcoal
- Catalytic oxidation

The following table is a summary of recommended conditions for controlled atmosphere storage. Only fruits and vegetables for which commercial uses of C.A. storage are common have been included.

	Temperature range		$\%O_2$ range	$\%CO_2$ range
	°C	°F		
Strawberry	0-5	32-41	10	15-20
Apple	0-5	32-41	2-3	1-2
Kiwifruit	0-5	32-41	2	5
Nuts and dried fruits	0-25	32-77	0-1	0-100
Bananas	12-15	54-59	2-5	2-5
Cantaloupe	3-7	38-45	3-5	10-15
Lettuce	0-5	32-41	2-5	0
Tomatoes				
Mature green	12-20	54-68	3-5	0
Partially-ripe	8- 12	47-54	3-5	0

llustrated here is a model of a small-scale commercial C.A. storage room. Each component is labelled. For farther information on construction methods, materials, and costs, please refer to the source listed below.

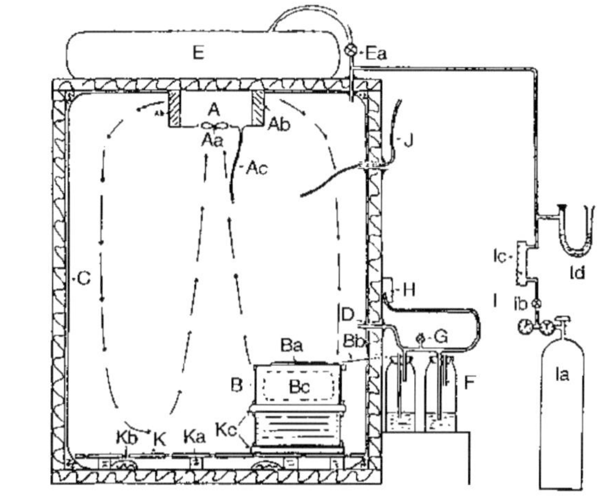

Controlled atmosphere storage of pallet loads of produce is also possible using a more permanent set-up for creating a gas-seal. Any number of pallets can be accomodated inside a plastic tent. A small trough constructed of sheet metal is laid in a rectangular pattern into a concrete floor of a storage structure. A 7-mil polyethylene sheet is put over the pallet load of produce, and the sheet is sealed by pushing a long piece of rubber tubing into the trough.

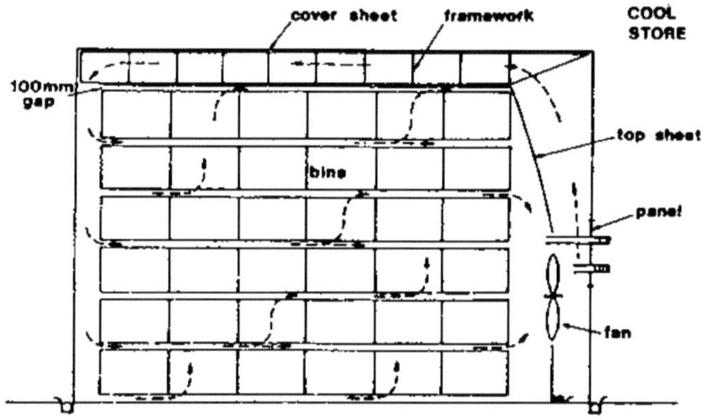

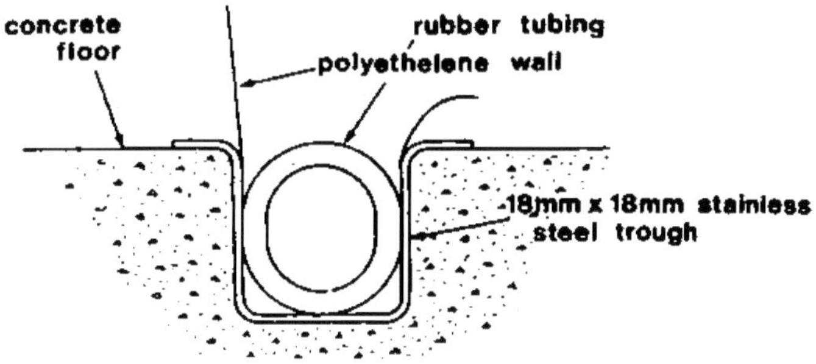

A low cost plastic tent fashioned from clear polyethylene sheeting can be used for controlled atmosphere storage of bunches of green bananas. A small fan serves to circulate the C.A. storage air (2% O_2 and 5% CO_2) through a chamber of potassium permanganate on aluminum oxide (Purafil). Ripening is delayed as ethylene is scrubbed from the storage air. The shelf life of bananas under these conditions is four to six weeks at ambient temperatures.

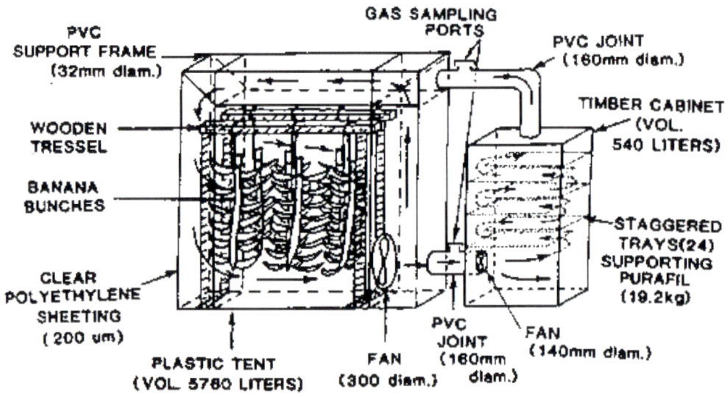

Relative Perishability and storage Life of Fresh Crops

Classification of fresh Crops according to their relative perishability and potential storage life in air at near optimum temperature and relative humidity.

Relative Perishability	Potential Storage Life (Weeks)	Commodities
Very high	<2	Apricot, blackberry, blueberry, cherry, fig, raspberry, strawberry; asparagus, bean sprouts, broccoli, cauliflower, green onion, leaf lettuce, mushroom, muskmelon, pea, spinach, sweet corn, tomato (ripe); most cut flowers and foliage; minimally processed fruits and vegetables.
High	2-4	Avocado, banana, grape (without SO_2 treatment), guava, loquat, mandarin, mango, melons (honeydew, crenshaw, Persian), nectarine, papaya, peach, plum; artichoke, green beans, Brussels sprouts, cabbage, celery, eggplant, head lettuce, okra, pepper, summer squash, tomato (partially ripe).
Moderate	4- 8	Apple and pear (some cultivars), grape (SO_2-treated), orange, grapefruit, lime, kiwifruit, persimmon, pomegranate; table beet, carrot, radish, potato (immature).
Low	8-16	Apple and pear (some cultivars), lemon; potato (mature), dry onion, garlic, pumpkin, winter squash, sweet potato, taro, yam; bulbs and other propagules of ornamental plants.
Very low	>16	Tree nuts, dried fruits and vegetables

Barn

Barn, in agriculture is a farm building for sheltering animals, their feed and other supplies, farm machinery, and farm products. Barns are named according to their purpose, as hog barns, dairy barns, tobacco barns, and tractor barns. The principal type in the United States is the general-purpose barn, used for housing horses and mules, cows, calves, and sheep and for storing hay and grain. Although the need for the general barn declined with the advent of tractors and electrical service, one or more barns are still found on the majority of North American and European farms. Many have been adapted to other uses.

Wood is the traditional barn material, but sheet steel and aluminum have been increasingly used since World War II, particularly on large farms. Barns usually consist of two stories, the first to shelter animals or machines and the second to store hay or grain, though one-story barns were gaining in popularity in the late 20th century.

Construction

The skeleton of a post and beam horse barn just after raising

Thomas Ranck Round Barn in Fayette County, Indiana, U.S.

In the U.S., older barns were built from timbers hewn from trees on the farm and built as a log crib barn or timber frame, although stone barns were sometimes built in areas where stone was a cheaper building material. In the mid to late 19th century in the U.S. barn framing methods began to shift away from traditional timber framing to "truss framed" or "plank framed" buildings. Truss or plank framed barns reduced the number of timbers instead using dimensional lumber for the rafters, joists, and sometimes the trusses. The joints began to become bolted or nailed instead of being mortised and tenoned. The inventor and patentee of the Jennings Barn claimed his design used less lumber, less work, less time, and less cost to build and were durable and provided more room for hay storage. Mechanization on the farm, better transportation infrastructure, and new technology like a hay fork mounted on a track contributed to a need for larger, more open barns, sawmills using steam power could produce smaller pieces of lumber affordably, and machine cut nails were much less expensive than hand-made (wrought) nails. Concrete block began to be used for barns in the early 20th century in the U.S.

Modern barns are more typically steel buildings. From about 1900 to 1940, many large dairy barns were built in northern USA. These commonly have gambrel or hip roofs to maximize the size of the hay loft above the dairy roof, and have become associated in the popular image of a dairy farm. The barns that were common to the wheatbelt held large numbers of pulling horses such as Clydesdales or Percherons. These large wooden barns, especially when filled with hay, could make spectacular fires that were usually total losses for the farmers. With the advent of balers it became possible to store hay and straw outdoors in stacks surrounded by a plowed fireguard. Many barns in the northern United States are painted barn red with a white trim. One possible reason for this is that ferric oxide, which is used to create red paint, was the cheapest and most readily available chemical for farmers in New England and nearby areas. Another possible reason is that ferric oxide acts a preservative and so painting a barn with it would help to protect the structure.

With the popularity of tractors following World War II many barns were taken down or replaced with modern Quonset huts made of plywood or galvanized steel. Beef ranches and dairies began building smaller loftless barns often of Quonset huts or of steel walls on a treated wood frame (old telephone or power poles). By the 1960s it was found that cattle receive sufficient shelter from trees or wind fences (usually wooden slabs 20% open).

Half-timbered barn with brick infill. Uetersen, Germany. This barn's proportions resemble a Low German house.

Half-timbered with wattle-work walls for ventilation. Stryd Lydan Barn, originally at Llannerch Banna, Flintshire, North Wales. Re-erected at the St Fagans National History Museum, Cardiff, Wales in 1951.

Wattle work walls in a sheep barn in Ruurlo, Netherlands.

Uses

The *hay track* developed in the early 19th century, here showing how the *hay hood* (roof extension) covers the track. The gable wall of this barn is missing.

In older style North American barns, the upper area was used to store hay and sometimes grain. This is called the mow (rhymes with cow) or the hayloft. A large door at the top of the ends of the barn could be opened up so that hay could be put in the loft. The hay was hoisted into the barn by a system containing pulleys and a trolley that ran along a track attached to the top ridge of the barn. Trap doors in the floor allowed animal feed to be dropped into the mangers for the animals.

In New England it is common to find barns attached to the main farmhouse (connected farm architecture), allowing for chores to be done while sheltering the worker from the weather.

In the middle of the twentieth century the large broad roof of barns were sometimes painted with slogans in the United States. Most common of these were the 900 barns painted with ads for Rock City.

In the past barns were often used for communal gatherings, such as barn dances.

Features

A barn in southern Ontario, Canada

A farm may have buildings of varying shapes and sizes used to shelter large and small animals and other uses. The enclosed pens used to shelter large animals are called stalls and may be located in the cellar or on the main level depending in the type of barn. Other common areas, or features, of an American barn include:

- A tack room (where bridles, saddles, etc. are kept), often set up as a breakroom.
- A feed room, where animal feed is stored – not typically part of a modern barn where feed bales are piled in a stackyard.
- A drive bay, a wide corridor for animals or machinery.
- A silo where fermented grain or hay (called ensilage or haylage) is stored.
- A milkhouse for dairy barns; an attached structure where the milk is collected and stored prior to shipment.
- A grain (soy, corn, etc.) bin for dairy barns, found in the mow and usually made of wood with a chute to the ground floor providing access to the grain, making it easier to feed the cows.
- Modern barns often contain an indoor corral with a squeeze chute for providing veterinary treatment to sick animals.

A barn of the Uster castle in the city of Uster, Switzerland

This barn in Thuringia, Germany has two *outshots* forming the recess to the middle barn doors.

Derivatives

The physics term "barn", which is a subatomic unit of area, 10^{-28} m^2, came from experiments with uranium nuclei during World War II, wherein they were described colloquially as "big as a barn", with the measurement officially adopted to maintain security around nuclear weapons research.

Types

Barns have been classified by their function, structure, location, or other features. Sometimes the same building falls into multiple categories:

- Apple barn or fruit barn: For the storage of fruit crops.
- Bank barn: A multilevel building built into a banking so the upper floor is accessible to a wagon, sometimes accessed by a bridge or ramp.
- Bridge barn or covered bridge barn: general terms for barns accessed by a bridge rather than a ramp.
- Boô: A sheep-barn and dwelling in the Netherlands, seasonal or sometimes year round.
- Pennsylvania barn (U.S.) of which there are sub-categories such as *standard* and *sweitzer* types. Also known as *forebay* or *porch barns*.
- Cantilever barn: A type of log crib barn with cantilevered upper floor(s) which developed in Appalachia (U.S.A.)
- Combination barn: Found throughout England, especially in areas of pastoral farming and the standard barn type in America. This general term means the barns were used for both crop storage and as a byre to house animals.
- Crib barn: Horizontal log structures with up to four cribs (assemblies of crossing timbers) found primarily in the southern U.S.A.
- Drying barns for drying crops in Finland and Sweden are called *riihi* and *ria*, respectively.
- New World Dutch Barn: An ancient barn type in the U.S.
- New England barn: A common style of barn found in rural New England and in the U.S.

- English barn (U.S.), also called a Yankee or Connecticut barn: An ancient and widespread barn type in the U.S.
- Granary: To store grain after it is threshed, some barns contain a room called a granary, some barns like a rice barn blur the line between a barn and granary.
- Gothic arch barn, has profile shaped as a Gothic arch, which became feasible to be formed by laminated members
- Housebarn, also called a byre-dwelling: A combined living space and barn, relatively common in old Europe but rare in North America. Also, longhouses were housebarns.
- Pole barn: A simple structure that consists of poles embedded in the ground to support a roof, with or without exterior walls. The pole barn lacks a conventional foundation, thus greatly reducing construction costs. Traditionally used to house livestock, hay or equipment.
- Potato barn or potato house: A semi-subterranean or two story building for storage of potatoes or sweet potatoes.
- Prairie barn: A general term for barns in the Western U.S.
- Rice barn and the related winnowing barn
- Round barn, built in a round shape the term often is generalized to the include polygonal barn and octagonal barn
- Swing beam barn: A rare barn type in part of the U.S. designed for threshing with animals walking around a pole held by a *swing beam* inside the barn.
- Tobacco barn: For drying of tobacco leaves
- Tithe barn: A type of barn used in much of northern Europe in the Middle Ages for storing the tithes: a tenth of the farm's produce which had to be given to the church
- Threshing barn: Built with a threshing floor for the processing and storage of cereals, to keep them in dry conditions. Characterised by large double doors in the centre of one side, a smaller one on the other, and storage for cereal harvest or unprocessed on either side. In England the grain was beaten from the crop by flails and then separated from the husks by winnowing between these doors. The design of these typically remained unchanged between the 12th and 19th centuries. The large doors allow for a horse wagon to be driven through; the smaller ones allow for the sorting of sheep and other stock in the spring and summer.

Other Farm Buildings Often Associated with Barns

- Carriage house: Cart shed
- Dutch barn (U.K.): An open sided structure for hay storage. The type with a movable roof is called a hay barrack in the U.S or a *hooiberg* (*kapberg*) in the Netherlands.
- A corn crib: A well ventilated storage space for dried ears of maize (corn).
- A granary or hórreo: A storage space for threshed grains, sometimes within a barn or as a separate building.

- Linhay (linny, linney, linnies): A shed, often with a lean-to roof but may be a circular linhay to store hay on the first floor with either cattle on the ground floor (cattle linhay), or farm machinery (cart linhay). Characterised by an open front with regularly spaced posts or pillars.
- Milk room or milk house: To store milk.
- Oast houses: An outbuilding used for drying hops as part of the brewing process.
- Shelter sheds: Open-fronted structures for stock
- Shippon: A shed which houses oxen and cattle. Has fodder storage above, regularly spaced doors on the yard side, a pitching door or window on the first floor.
- Stable: Usually for housing horses.

Temperature Control with Respect to Different Fruits/Vegetables/Grains

Whether you are a grower or consumer of fruits and vegetables, at some point you will probably be concerned about the optimal temperature at which to store your produce. Since we are in the cold storage business, we thought a handy guide would be fun to share with our readers.

There are many benefits of keeping your vegetables and fruits at the optimal temperature until eating or cooking with them. Produce that is chilled appropriately will always taste better than produce kept at warmer temperatures where spoiling could occur. Also, as you might expect, produce will last longer when stored at the proper climate. Finally, storing fruit and vegetables at the right temperature helps prevent the slow growth of bacteria that can lead to foodborne illnesses.

Below, we have grouped various fruits and vegetables by the optimal temperature at which each item should be stored, from coldest to warmest. For cold storage produce, it is recommended to store the fruits or vegetables pretty close to the optimal temperature (within 1-2 degrees). As you increase in optimal storage temperature, a wider temperature range is more acceptable (5-7 degrees). All temperatures are Fahrenheit.

Just Below Freezing:

Pears	(30°)
sweet cherries	(30.5°)
horseradish	(31°)
apricots	(31.5°)
currants	(31.5°)
elderberries	(31.5°)
Gooseberries	(31.5°)
Grapes	(31.5°)

Jerusalem artichokes	(31.5°)
Nectarines	(31.5°)
Plums	(31.5°)
Prunes	(31.5°)
quinces	(31.5°)
raspberries	(31.5°)

Freezing:

Bean sprouts	(32°)
broccoli	(32°)
brussels sprouts	(32°)
bunched or topped beets	(32°)
cabbage	(32°)
carrots	(32°)
cauliflower	(32°)
celery	(32°)
celeriac	(32°)
chard	(32°)
collards	(32°)
endive	(32°)
escarole	(32°)
garlic	(32°)
green peas	(32°)
greens	(32°)
kale	(32°)
kohlrabi	(32°)
leeks	(32°)
lettuce	(32°)
mushrooms	(32°)
parsley	(32°)
parsnips	(32°)
radishes	(32°)
rhubarb	(32°)
rutabagas	(32°)
salsify	(32°)
sour cherries	(32°)
spinach	(32°)
strawberries	(32°)
sweet corn	(32°)
turnips	(32°)
watercress	(32°)
witloof chicory	(32°)

Just Above Freezing:

Blackberries	(32.5°)
artichokes	(33.5°)
asparagus	(33.5°)
beets	(33.5°)
blueberries	(33.5°)
bok choy	(33.5°)
cauliflower	(33.5°)
cherries	(33.5°)
chicory	(33.5°)
Chinese pea pods	(33.5°)
figs	(33.5°)
green peas	(33.5°)
herbs	(33.5°)
kiwi	(33.5°)
napa cabbage	(33.5°)
onions	(33.5°)
persimmons	(33.5°)
salad	(33.5°)
sprouts	(33.5°)
tangerines	(33.5°)
pre-cut fruit	(34°)
pre-cut vegetables	(34°)
apples	(35°)
cantaloupe	(37°)
lima beans	(39°)

40-49 Degrees:

Cranberries	(40°)
southern peas	(40.5°)
hot chili peppers	(41°)
ripe avocado	(41.5°)
green beans	(42.5°)
lychees	(42.5°)
oranges	(42.5°)
snap beans	(42.5°)
pomegranates	(45.5°)
summer squash	(45.5°)
guavas	(47.5°)
okra	(47.5°)
potatoes	(47.5°)
unripe avocado	(47.5°)

50-59 Degrees:

Eggplant	(50°)
sweet peppers	(50°)
winter squash	(50°)
limes	(51.5°)
cucumbers	(52.5°)
mangos	(52.5°)
melons	(52.5°)
papaya	(52.5°)
pineapple	(52.5°)
pumpkin	(52.5°)
lemons	(53.5°)
basil	(54.5°)
dry beans	(55°)
coconuts	(57.5°)
grapefruit	(57.5°)
sweet potatoes	(57.5°)
ripe bananas	(58°)

60-69 Degrees:

Jicama	(60°)
ginger root	(62.5°)
tomatoes	(62.5°)
watermelon	(62.5°)
green bananas	(66°)

If your farm's produce distribution plan requires the cold storage of fruits or vegetables in the Seattle area, don't hesitate to contact Washington Cold Storage.

The most influential factors in the storage of grains is moisture and temperature. Grains containing less than 12% moisture and pulses (legumes harvested solely for the dry seed) with less than 10% moisture can be stored for food purposes indefinitely. Grains include barley, corn, millet, oats, rice, rye, sorghum, triticale and wheat. Pulses include beans, broadbeans (fava beans), chickpeas, lentils and dried peas. High moisture content in grain of over 12% causes damage to the seeds because it promotes diseases. At 13 ½ to 15% moisture levels some fungal spores begin to grow, other species of fungi require 16 – 23%. Aerobic bacteria require 20% to grow. Grains stored with these moisture levels will spoil and become unfit for food and can spontaneously ignite under ideal conditions. The best temperature for storage of grains is between 40 – 60 degrees. Higher temperature affects the ability of the stored seeds to germinate over time, but food value is only slightly reduced. Freezing temperatures does not damage stored grains or pulses.

Cold Storage

Cold storage facilities are crucial to minimize post-harvest losses; however, losses occur at every step in the post harvest cycle, and therefore cold storages cannot be considered as independent solutions to prevent post harvest spoilage but as one component that needs to be integrated in a cold chain network from the point of harvest to the point of purchase by the end consumer

The different stages in the post-harvest cycle can be summarized as:

- Harvest
- Precooling
- Transportation & packaging
- Cold storage
- Display at the market

During these different stages, not only temperature control, but also proper and minimal handling of produce including handling, cleaning, sorting and adequate packing is crucial.

Additionally, socioeconomic factors add to the post harvest loss issue: the lack of marketing systems, transportation facilities and market information are key hindering factors. Also, government regulation, often intended to keep prices low for consumers or to ensure a certain minimum price for farmers, can be counterproductive regarding incentivizing the production of high quality produce. Often, simply the required equipment, such as harvesting tools, containers, cleaning, packing and cooling equipment are not available. Furthermore, the knowledge on post harvest handling and marketing of farmers and other actors handling the produce, for example during transport, is generally very limited leading to suboptimal treatment of the produce.

Harvest, Handling and Transport

Adequate and proper harvesting and handling of fresh produce are crucial preconditions for long storage and shelf lives of fruit and vegetables. Considering the harvesting and subsequent handling stage of produce, the thumb rule of, the less the better should be followed, in order to minimize chances of mechanical injuries of produce. Once damaged, shelf life is considerably compromised.

Harvesting principles

For harvest procedures the subsequent points are important to consider:

- Choosing the correct harvest time refers to both the ripeness and maturity of the produce as well as the right time of the day. Optimal harvest times for most produce is either early morning hours or the evening when temperatures are lower
- Harvesting of delicate, high value produce is best done manually (especially when labor costs are low and fuel costs are high

- Do not place produce on the ground directly, but use harvesting mats or containers/baskets instead

Handling

Considering the handling of harvested produce, the following points are important:

- Before being put into storage, produce should be sorted and graded with regards to quality (only high quality produce should enter the storage facility)
- Produce needs to be cleaned (with clean water in order to avoid the spread of molds and fungi) before being put into storage containers and entering the storage rooms. Dirt bears the potential of introducing pests into the storage facility
- Time span between harvest and the placement into storage needs to be kept as short as possible

Transport

For the transport of fresh produce, the following points need to be considered:

- Trucks, carts, etc. should not be overloaded to avoid damage
- Use high quality packaging that will withstand the transport

Temperature Control - Cold Storage

Fresh food products are characterized by their perishability, and levels of vulnerability to pests and spoilage; however, in general the control of temperature and relative humidity and the prevention of damage can increase shelflife, especially of fresh vegetables and fruits, tremendously. Biological deterioration, caused by respiration rate, ethylene production, mechanical injuries, water stress, physiological disorders and pathological breakdown, leads to decay, loss in nutritive value, and changes in color, texture and flavor. Factors influencing the rate of deterioration are temperature, the level of relative humidity, air velocity, atmospheric composition, i.e. concentration of oxygen, carbon dioxide and ethylene, as well as sanitation.

Molding, pests and other spoilage reasons which lead to quality deterioration, such as loss of water, thrive under warm and humid conditions – conditions which can be commonly found in India. Most factors causing product deterioration can be addressed through temperature control, i.e. cooling. A reduction in temperature lowers rates of physiological change (respiration, ethylene production, and enzymatic processes) and slows down the growth of microorganisms (bacteria and fungi). Produce after the point of harvest is still 'alive' meaning that processes of the metabolism, such as respiration, as well as the activity of microorganisms remain ongoing leading to product deterioration. In order to lower respiration rates, delay ripening and to reduce water loss temperature needs to be controlled, either by avoiding the exposure of heat or by creating cooling conditions thereby increasing a product's shelf life. Adequate storage facilities can, to a certain extent, control factors such as temperature, relative humidity and air velocity, increasing product quality, shelf-life and value.

The table below demonstrates the effect of temperate control on shelf life of fresh produce. Temperature reduction, even if kept above the level of optimum storage conditions, can yield tremendous extensions of storage life. Accordingly, any form of cooling, such as moving produce to a shaded and ventilated area or choosing harvest hours in the early morning or evening, is beneficial.

Temperature (°C)	Relative Rate of Deterioration	Relative Shelf Life	Example of Potential Shelf Life in Days
0	1.0	100	45
10	3.0	33	15
20	7.5	13	5
30	15.0	7	2.5
40	22.5	4	1.3

Effect of temperature control on shelf life of fresh agricultural products

Commodity	Recommended Temperature for Handling and Storage (max post-harvest life)	Post-harvest Life at 35°C* (ambient temperature)	Post-harvest Life at 25°C	Post-harvest Life at 15°C	Increased Marketing Time Available at 15°C
Cabbage	0°C (6 months)	2 weeks	4 weeks	8 weeks	4X
Carrots	0°C (6 months)	2 weeks	4 weeks	8 weeks	4X
Tomatoes	15°C (14 days)	3 days	6 days	14 days	5X
Peppers	12°C (20 days)	3 days	7 days	15 days	5X
Potatoes	5° to 7°C (5 to 10 months)	2 weeks	4 weeks	8 to 10 weeks	4X
Spinach	0°C (14 days)	1 day	2 days	5 days	5X
Sweet potatoes	15°C (4 to 6 months)	1 month	2 months	4 to 6 months	4X

Effect of temperature control on shelflife of fresh agricultural products

After produce has been pre-cooled, it needs to be transported and put into a cold storage facility where, in order to maximize shelf-life and quality, it should be stored at its optimum storage temperature which varies according to different items making the storage of several items in one facility problematic if multiple storage chambers are not available. Storage and shelf life depends on intrinsic characteristic of items stored as well as the storage conditions.

In general, the optimum storage temperature is as low as possible but still above the level of chilling sensitivity of the respective crop. Another important factor affecting shelf-life of fresh produce is the level of relative humidity, impacting water loss, decay, physiological disorders, and the uniformity of ripening. Relative humidity levels should be high and saturated for most types of produce. Finally, ethylene also impacts shelf life: whilst some items are ethylene producers after harvest, other crops are sensitive to ethylene exposure. Subsequently, if produce cannot be stored separately, their requirements regarding temperature, relative humidity and ethylene production/ exposure need to be considered.

According to the Postharvest Centre of California, for the sake of short-term storage or transport, three groups of produce can be kept together:

- Most leafy vegetables, cole crops, fruits and berries from temperate climates:0°-2° C, 90%-98% relative humidity.
- Citrus fruits, most tropical fruits, many fruit-type vegetables: 7°-10° C, 85%-95% relative humidity.

- Root-type vegetables, winter squashes, most tropical fruits and melons: 13°-18° C, 85%-95% relative humidity.

Cold Storage Technologies for Small-scale Production

Cooling and cold storage are very energy intensive and expensive undertakings that require reliable electricity supply. This precondition is often not given and capital and running costs for farmers in developing countries are generally too high. Hence, unbroken cold chains and the use of appropriate cold storage facilities are generally not accessible for small-scale farmers .

Below, several low technology options are being presented that can operate in electricity scarce regions and are feasible for small-scale farming operations:

Technology	Description	Temperature provided	Fuel
passive/evapaorative cooler	works in naturally dry areas without the input of energy	10-25° C	• no fuel • nocturnal cooling • architectural measures (shade creation, fountains, etc.)
absorption refrigerator	heat-driven technology	below 10° C	• solar • kerosene
refrigerators	electricity-driven and therefore dependent on reliable electricity supply	0° C	• grid • diesel • renewable sources (hydro, solar, batteries, etc.)

There are multiple factors to consider in the technology and site selection of a cold storage facility:

- Location: the location chosen should be easily accessible and close to market hubs and/or produce collection points.
- Electricity: the access to and the supply of electricity to the cold storage facility should be given. Electricity supply has an impact on the size and equipment feasible of the cold storage.
- Drainage and waste disposal: good drainage options and waste management equipment needs to be readily available.
- Water: supply of water is important for matter of sanitation. Also the removal of wastewater generated has to be given.
- Economic viability: the value generated though cold storage needs to justify investment and running costs.

In general, choosing an appropriate cold storage technology depends on the temperatures and level of reliability required, resources available – including skilled labor, capital, electricity and water supply

Cold Storage Management

Temperature control can have huge benefits for the improvement of shelf-life and product;

however, benefits will only realize if the cold storage is managed properly with regards to temperature, relative humidity levels, air circulation, adequate space between storage bins, trays and containers, the mixing of compatible produce, as well as the management of product in and outflow which should follow the 'First In, First Out' principle.

The following points should therefore be considered:

- Temperature management

According to information of the FAO, cold storage facilities should be cooled to the storage temperature required at least three days before produce is being entered.

Chilling Injury

Some produce, especially of tropical origin, is sensitive to chilling, which means that it will incur physiological damages if stored at a certain time period below a certain temperature but above their freezing points. In general, the longer the time period that produce is exposed to temperature below their level of chilling sensitivity and the lower the temperature, the faster damages will occur. It should also be noted that effects can be of a cumulative nature, i.e. the time periods of storage below the level of chilling sensitivity add up even if produce is stored at optimal conditions in between. Several factors, such as the level of maturity and level of ripeness at the point of harvest can affect chilling sensitivity .

Common visual symptoms:

- Surface lesions (pitting, large sunken areas and discoloration).
- Water soaking (disruption of cell structure and accompanying release of substrate favors the growth of microorganisms).
- Internal discoloration of the pulp.
- Failure to ripen in the expected pattern.
- Accelerated rate of senescence.
- Increased susceptibility to decay.
- Compositional changes.

Freezing Injury

Below 0°C all type of produce freezes due to dissolved soluble solids which are present in cell saps. Often, damages incurred through freezing only become visible once the produce is returned to temperatures above 0° C.

Symptoms:

- Waterlogging
- Glassy areas in the flesh

Storage Mix

Different produce has different storage demands with regards to factors, such as optimum temperature, level of relative humidity, levels of ethylene production and sensitivity, and chilling sensitivity. Hence, the storage of single commodities is less complex than the storage of multiple commodities; however, the latter can still be the more viable option depending on external circumstances. Besides damages due to mismatches in temperature, also odors can be transmitted and ripening and decay can be affected through the exposure to ethylene, which can result in changes in color, flavor and texture. Further, only high quality produce should be 'allowed' in the cold storage and produce should be sorted accordingly

Examples of odour transfers which should be avoided:

- Apples/pears with celery, cabbage, carrots, potatoes or onions
- Celery with onions or carrots
- Citrus with strongly scented vegetables
- Pears/apples with potatoes à former acquire unpleasant taste
- Green pepper will taint pineapples
- Onions, nuts, citrus, potatoes should be stored separately

Examples of ethylene producing and ethylene sensitive products:

- Ethylene producing: e.g. apples, avocado, bananas, pears, peaches, plums, tomatoes
- Ethylene sensitive produce: e.g. lettuce, cucumbers, carrots, potatoes, sweet potatoes

Stacking

Adequate airflow and the even distribution of cold air need to be ensured. It is important to note that air always takes the path where resistance is lowest, and hence partly or unevenly filled storage facilities will have poor cooling rates. The following points should be considered:

- A gap of at least 8 cm between walls and the floor, and the stacks of produce should be kept in order to ensure air flow
- Well-ventilated storage boxes/containers/crates will improve cooling speed, such as PVC crates or ventilated boxes made of cardboard
- Adequate space in between storage pallets should be about 4-6 inches

Operation and Maintenance

Cold storage is a very costly and energy intensive undertaking and therefore the door to the cold storage should be opened as few times as possible and that also in the early morning or evening hours. Also the usage of lighting and fans should be kept at a minimum. Further points should further be considered:

- Logbook: operators should keep track of incoming and outgoing, as well as produced stored, including anticipated arrival time, quantity, quality, anticipated storage time, storage conditions required, sales price on day stored, actual sales price, energy consumed, etc. of new produce .

- Inventory: an up-to-date inventory should be kept including information about location within cold storage, type, quantity and harvest date of produce, use of pre-cooling method, date produce entered the facility, etc.
- Loading: in general, loading rates should be 4-5% of total capacity. The cold storage should be divided in areas for short-term and longer-term storage space

Precooling

Pre-cooling is done to remove the field heat of the harvested produce, which is detrimental to keeping quality of fruits and vegetables and it is done to retard ripening and senescence processes. Prompt pre-cooling conserves the weight and extends the storage life in tomato. PLW in storage can be reduced from 6 to 2.5% by employing pre-cooling treatments.

- Methods of Pre-Cooling:

  Several effective methods for rapid removal of heat from produce are in commercial use. The choice of method depends largely on the perishability and refrigeration equipment of the produce its adaptability to a specific method and the availability of facilities.

- Hydro Cooling:

  Cooling with cold water is rapid and effective method of pre-cooling used for cooling a wide range of fruits and vegetable in bulk before packing. Its use is limited for packed commodities because of the difficulty of achieving sufficient water flow through the containers. Flooding, spraying or immersion accomplishes hydro cooling. A properly designed flood system is more efficient than either the spray or immersion system because it combines a great volume of water with rapid movement of cooling medium over the product. When cooling completed, the product must be moved to a cold room or raid reworking will occur.

- Air –cooling or Room Cooling:

  The use of refrigerated air as pre-cooling medium is widely used for pre-cooling packed fruit, but the system is not widely used for vegetables. Pre-cooling with air can be accomplished in a conventional cold storage room, a special pre-cooling , a funnel cooler, or a forced air cooler. Cooling with air requires a longer time than cooling with water or vacuum.

- Vacuum Cooling:

  Leafy vegetables are commonly cooled by reducing atmospheric pressure in artificial hermetically sealed chambers. Reducing the atmosphere pressure also reduces the pressure or water vapour in the chamber and thus cooling is effected. The outstanding advantages of vacuum cooling are the sped and uniformity of cooling of adapted commodities. Leafy vegetables , particularly lettuce is difficult to cool with water or air, but they can be field packed and then cooled quickly and uniformity by vacuum. Commodities like tomatoes with epidermis which is resistant to water movement are not adopted to vacuum cooling.

- Pre-cooling of Fruits:

  Temperature protection of the fruit in the field immediately after harvesting is very important. This can easily be done by moving the fruits to the shade. It is essential to remove the field- heat of fruits in order to have extended shelf life.

Following Methods are Adopted to Pre-cooling the Fruits:

- Air Cooling:

  The air cooling is done by placing the fruits in the cold room. Fruits are placed in well ventilated containers in order to achieve some air exchange.

- Hydro-cooling:

  The hydro-cooling is an old and effective pre-cooling method for fruits. Fruits are dipped in cold water or spray the cold water on the fruits. Some chemicals are also mixed with water used in hydro-cooling to prevent the shade and diseases.

  Hydro-cooling appears with a refrigerated $CaCl_2$ solution (21 °C for 10 minutes) would prevent storage disorders. Pre-cooling has got effect on quality of fruits e.g. L-49 variety of guava. The reducing sugar percentage and ascorbic acid content was higher in pre-cooled fruits than without pre-cooled fruits.

- Vacuum Cooling:

  It is costliest method of pre-cooling. It is rarely used in case of fruits. It is done by reducing the atmospheric pressure. It reduces the pressure of water vapour in chamber which result in evaporation of water from fruits which bring down the temperatures.

  Pre-cooling treatment prior to low temperature storage can delay the development of chilling injury in grapes.

Pre-cooling of Vegetables:

When the vegetables are harvesting during hot weather, it is desirable to remove the filed heat of the harvested vegetables to retard the ripening and senescence. Cooling and storage losses can be reduced by pre-cooling when the vegetable harvested during hot period pre-cooling is of vital importance. Eg. Melons.

Stages of ripening a level of field heat and the vegetable also determines the need of processing. In case of Tomato above 26.7° C temperatures and ripening is to delayed there is no need for pre-cooling.

Pre- cooling can be done by Following Processes:

- Placing the vegetables in a refrigerated truck with forced humidified air circulation.
- Placing ice in package.
- Placing ice in water and passing the vegetables through a spray of cold water.
- Vacuum cooling is also done.

- Hydro-cooling is done by flooding spraying or immersing is a rapid and effective method of pre-cooling.

Water is an excellent material to transfer a heat from the produce to the cooling medium.

Highly perishable leafy vegetables and sealed crops are often cooled by reducing the atmospheric pressure. Leafy vegetables are especially difficult to cool with and then cooled rapidly and uniformity by vacuum cooling.

Asparagus cooled in water 0.5 to 1.0° C for 9 to 12 minutes and then stored at 2° C for 1-5 days followed by 2 days at 20° C reduces water losses and quality reduction during storage.

Capsicum harvested at green when cooled in water at 7 to 10° C for 10 minutes and then packed in poly bags with or without ventilation shows largest shelf-life than non-pre cooled fruits.

Hydro cooling increases the storage life of melons when stored in 21 % O_2 + 0% CO_2.

Pre-cooling benefits include:

- Lowering the required workload of a cold storage since optimum storage temperature is reached more quickly.
- Restricting and minimizing respiratory activity, thereby conserving the weight of the produce, and enzymatic degradation of the produce harvested; thus preventing softening, water loss and wilting.
- Preventing microbial growth, such as bacteria and fungi thereby decreasing the rate of decay.
- Decreasing rate of ethylene production and the impact on ethylene sensitive produce.
- Delaying chilling injuries for certain fruits.
- Increasing the daily intake into storage facilities which should not exceed 10% of its cooling capacity if produce is not pre-cooled.

Hydrocooling

When warm produce is cooled directly by chilled water, the process is known as hydrocooling. Hydrocooling is an especially fast and effective way to cool produce. With modern technology, hydrocooling has now become a convenient and attractive method of postharvest cooling on a large scale.

Many types of produce respond well to hydrocooling. Produce items that have a large volume in relationship to their surface area (such as sweet corn, apples, cantaloupes, and peaches) and that are difficult to cool can be quickly and effectively hydrocooled. Unlike air cooling, no water is removed from the produce. In fact, slightly wilted produce may sometimes be revived by hydrocooling.

Hydrocooling is one of several postharvest cooling methods available to growers, packers, and shippers.

Hydrocooling:

- Cools produce rapidly (about 15 times faster than air).
- Allows for greater harvesting and marketing flexibility.
- Easily handles large amounts of produce (although hydrocoolers can also be designed for smaller quantities of produce).

There are also some limitations to using hydrocoolers:

- They can be used only for produce not sensitive to wetting. Many diseases are encouraged by wetting.
- Some hydrocoolers are not as energy efficient as other methods and therefore may not be as cost effective in some situations.
- There are some restrictions on the types of packaging and stacking arrangements used for produce that is to be hydrocooled.

Hydrocooling Methods

In most hydrocoolers, a pump moves chilled water into contact with warm produce. The warmed water is then recooled and recycled. For cooling the water, many hydrocoolers have a vapor-compression refrigeration system similar to an air conditioner or refrigerator. A refrigeration system can be thought of as a pump that moves heat. The capacity of a refrigeration system to move heat is measured in tons. One ton of refrigeration is equivalent to 12,000 Btu per hour.

Some hydrocoolers do not use a refrigeration system. Instead, crushed or chunk ice is used to cool the water. Typically, large blocks of ice weighing as much as 300 pounds are trucked from an ice plant, crushed, and added as needed to a water reservoir attached to the hydrocooler. The capital cost of a hydrocooler of this type is much less than one with an integral refrigeration system and may be preferred by growers with a limited amount of produce or a short cooling season. However, to make a valid economic comparison, the cost of the ice must be considered. For a hydrocooler of this type, a reliable source of ice must be available at a reasonable cost.

Produce Packaging and Stacking Considerations

The design of the produce package and the stacking arrangement is critical to the heat transfer process in hydrocooling. A variety of produce packages have been successfully used in hydrocooling. These packages include wire-bound wooden crates, waxed fiberboard cartons, mesh poly bags, and bulk bins. Palletized packages can be hydrocooled if they are carefully stacked to allow water to enter the packages. If the water flows around and not through the packages, little cooling will occur. Produce in waxed cardboard cartons with solid tops is particularly difficult to cool because the top does not allow water to enter.

Wire-bound cartons and crates with a large percentage of open space are more suitable for hydrocooling because they allow for sufficient entry of water. Produce in 20-bushel bulk bins cools especially well because the cool water can easily percolate down through the product.

Types of Hydrocoolers

There are several different hydrocooler designs. Hydrocooling methods differ in their cooling rates and overall process efficiencies. In addition, hydrocoolers vary in the method of cooling that is used and the method of moving or placing the produce so that water comes in contact with it. Four types of hydrocoolers that are commonly used are discussed below: conventional, batch, immersion, and truck hydrocoolers.

Conventional Hydrocooler

A conventional hydrocooler allows the produce, either in bulk bins or in cartons, to pass along a conveyor under a shower of chilled water. Warm produce is placed on one end of the conveyor, and cooled produce is removed at the opposite end. The rate at which the conveyor, and thus the produce, is advanced through the shower is about 1 feet per minute and may be varied on most hydrocoolers to suit conditions.

Manufacturers of conventional hydrocoolers often specify their different models by length. For example, a "10-feet" hydrocooler has an active cooling length of 10 feet, even though it may be as much as 20 feet long. The additional length is used for input and output conveyors. The longer the active cooling area, the greater the capacity of the hydrocooler. Conventional hydrocoolers with as much as 50 feet of active cooling length and a width of 8 feet have been constructed. However, specifying the capacity of a hydrocooler entirely by length can be misleading if conveyor speed, conveyor width, water temperature, and flow rate are not taken into consideration. Therefore, when comparing different hydrocoolers, buyers should find out how many pounds or bushels of a certain commodity the hydrocooler will cool per hour from one temperature to another. Most hydrocooler manufacturers can readily supply this type of data. Hydrocooling requires large quantities of water to be passed by the produce. Water flow rates as great as 20 gallons per minute per square foot of active cooling area are common. For example, a hydrocooler with an active cooling area 4 feet wide by 20 feet long (80 square feet) would require the circulation of 1,600 gallons per minute.

Most conventional hydrocoolers are high-production units with large refrigeration systems and heavy-duty components. Because of their relatively high cost, they must be operated for considerable periods each year to be economically justified. With the relatively small acreages of produce common in North Carolina, these hydrocoolers would need to be used by more than one grower (or with more than one crop) or by a co-op of growers and packers in order to be cost effective.

Batch Hydrocooler

Batch hydrocoolers are enclosures that do not have conveyors. Palletized cartons or bulk bins of produce are loaded into the enclosure with a fork lift. The door of the enclosure is then closed, and large quantities of chilled water are flooded over the top of the produce, collected at the bottom, recooled, and recycled.

Most batch hydrocoolers can cool only one pallet of produce at a time. However, some larger batch units are occasionally built that can cool as many as eight pallets at once. These hydrocoolers generally have a smaller capacity than conventional hydrocoolers and therefore may be less expensive. They are better suited to growers with a limited amount of produce that could not economically justify a larger unit.

A frequent complaint about both conventional and batch hydrocoolers is that they do not cool all containers uniformly. The chilled water may not be evenly distributed throughout the load, resulting in undercooling of some parts. To overcome this deficiency, some batch hydrocoolers use a high-capacity fan to pull a fine mist of chilled water through the produce packages. The forced air has the effect of making the cooling more consistent because it pulls the water past the produce more evenly than would occur by gravity flow alone. This design, known as "hydro-air-cooling," has been successfully used with items that are particularly difficult to cool.

Immersion Hydrocooler

Immersion hydrocoolers are large, shallow, rectangular tanks that hold moving chilled water. Crates or boxes of warm produce are loaded into one end of the tank and moved by a submerged conveyor to the other end where they are removed. Crushed ice or a vapor-compression refrigeration system keeps the water cold, and a pump keeps the water in motion. Most produce is only slightly buoyant so it tends to stay submerged. The length of time the produce remains in the water varies with the initial conditions and desired ending temperature.

Test have shown that the most rapid hydrocooling is obtained by immersing produce in tanks of agitated chilled water. Immersion hydrocooling is nearly twice as rapid as conventional hydrocooling methods. With conventional hydrocooling, the cold water that is sprayed or flooded over the produce contacts only a portion of its surface. The result is less than maximum heat transfer. Immersion hydrocooling reduces the temperature more rapidly because moving chilled water completely surrounds the exterior surface.

Immersion hydrocooling was once practiced by placing loose produce into tanks of chilled water. This method is seldom practiced today because it is labor intensive and much of the cooling effect is lost when the produce is packed. In addition, packing shed workers are often reluctant to handle the wet, cold produce. The bulk "fluming" of string beans in chilled water prior to grading and packing is an example of immersion hydrocooling. Bean fluming is gradually being replaced by other hydrocooling methods.

Truck Hydrocooler

Some sweet corn growers in North Carolina use a cooling method known as "truck hydrocooling." In this process, water is cooled and delivered to the produce by a simple system consisting of a tank holding several thousand gallons of water, a high-capacity pump, associated piping and valves, and an ice crusher. Graded and packaged produce (as much as 600 to 800 crates) is loaded from the field into a tractor trailer. Perforated pipes are then inserted into the trailer above the load to deliver a shower of chilled water from the stationary cooling system. The water flowing out of the trailer is collected, recooled, and recycled. Water flow rates of 1,000 gallons per minute for this type of hydrocooler are common. After an hour or more of hydrocooling, the pipes are withdrawn, a layer of crushed ice is added over the top of the crates, and the produce is immediately shipped.

To cool the water, blocks of ice from a commercial ice plant are crushed and added to the tank. Approximately one 300-pound block of ice is required per 100 gallons to cool the water initially. As the water warms during the cooling process, ice is added periodically.

Truck hydrocoolers can be built by the grower at a central location on his or her farm for a small fraction of the cost of a commercial hydrocooler. A truck hydrocooling system capable of cooling several truckloads of fresh produce per day can be built from locally available parts.

Until recently, it was assumed that truck hydrocooling removed 30 to 40 degrees of field heat from sweet corn cooled in this manner. However, recent tests conducted at several locations in eastern North Carolina have not supported this common belief. Temperature sensors positioned throughout a trailer load of sweet corn hydrocooled for 1 hour and 45 minutes showed an average decline in temperature of only 15° F. This poor cooling rate apparently occurs because most of the water flows through the spaces between the crates, decreasing the total heat transfer.

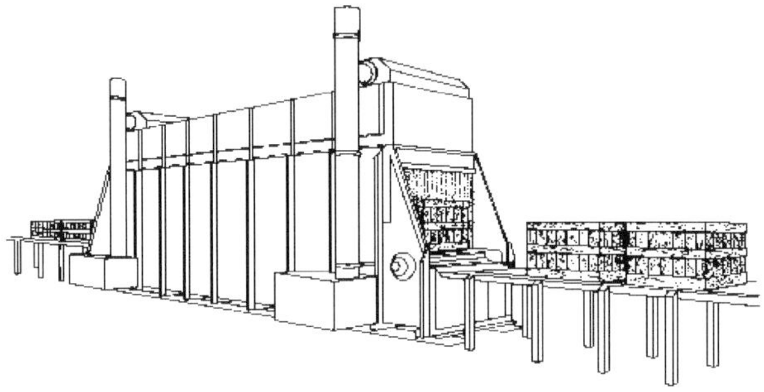

Figure: Typical conventional hydrocooler

Figure: Single pallet batch cooler

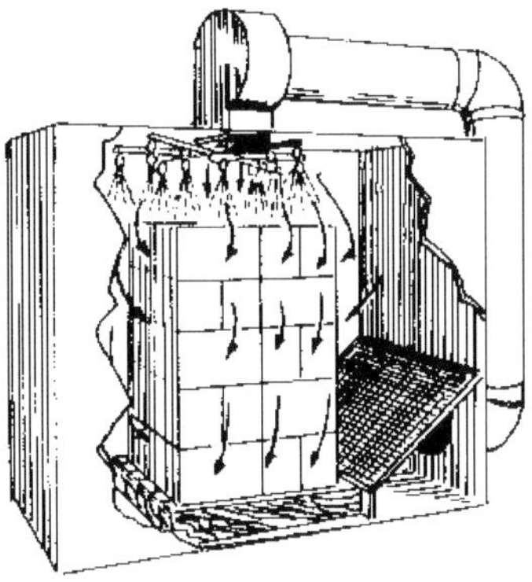

Figure: Cut-away view of a hydro-air-cooling hydrocooler

Figure: Immersion hydrocooler

Figure: Truck hydrocooling operation in Eastern North Carolina

Determining Hydrocooling Rates

To operate a hydrocooler effectively and efficiently, it is necessary to understand how water removes heat from the produce and the factors that affect the rate of cooling for various types of produce. For more precise estimates and information, consult a hydrocooler manufacturer or your county Extension agent.

Heat Transfer

When any object is warmed or cooled, the physical process is known as heat transfer. There are several different mechanisms of heat transfer. In the process of hydrocooling, the heat in the produce is carried away by the currents of moving water. This type of heat transfer is known as convection. Several factors contribute to the rate at which heat is transferred during convection. One of these factors, called the heat transfer coefficient, remains nearly constant for all types of produce. The heat transfer coefficient is a measure of how fast heat is transferred from the surface of an object. A commonly used value for the heat transfer coefficient is 120 Btu per hour per square foot per degree Fahrenheit.

The basic heat transfer equation governing convection is:

$$Q = h \times A \times (Ts - Tw)$$

where:

Q = the rate at which heat is transferred from a batch of produce, in units of Btu/hr

h = the heat transfer coefficient, in units of Btu/(hr ft 2° F)

A = the total exposed surface area of the produce, in square feet, over which the heat transfer occurs

Tw = the temperature of the water (°F)

Ts = the temperature of the produce surface (°F)

From the equation, it may be seen that the rate of heat transfer, Q, is directly proportional to the magnitude of the heat transfer coefficient, the amount of surface area, and the temperature difference between the surface of the produce and the water.

Other factors also affect the rate of heat transfer, including the surface area of the produce and the temperature difference between the surface of the produce and the chilled water. These factors relate directly to the time it takes to cool a specific type of produce. This means that the rate of heat transfer for produce increases when the surface area of the produce increases or when the difference between the produce temperature and the water temperature increases.

The surface area that is exposed to the chilled water varies considerably among different types of produce. For example, the surface area of 10 pounds of sweet corn is considerably less than that of 10 pounds of snap beans. Therefore, if hydrocooled in water of the same temperature, snap beans can be expected to cool much faster than sweet corn. In addition, the greater the difference between the water and the produce temperature, the faster the cooling.

So far, the discussion has centered on heat transfer from the standpoint of the surface temperature of produce. When cooling produce, the rate of cooling of the interior of the produce is of primary importance. The theoretical basis just reviewed will promote a better understanding of the factors that affect the hydrocooling rates of various produce and hydrocoolers.

Calculating Cooling Rates

Field and laboratory studies have provided useful data on hydrocooling rates for many fruits and vegetables. This information can be used to calculate how long it takes to cool the interior of a particular product to a desired temperature if the starting temperature and the water temperature are known.

Researchers have developed a graph summarizing current information about heat transfer for various fruits and vegetables. Presented in figure below, this graph shows a family of eight ideal cooling lines for various types of produce immersed in agitated chilled water.

Figure below KEY:

A – Greens

B – Beans, peas, asparagus

C – Small cucumbers, radishes, beets (< 1.5" dia.)

D – Small apples and peaches, slicing cucumbers

E – Sweet corn, apples, and peaches

F – Large apples and peaches (> 3" dia.)

G – Cantaloupes, large eggplant

These lines are approximations only and are based primarily on the physical size of the produce. The horizontal axis is the cooling time in minutes and the vertical axis is the decimal temperature difference (DTD) at the temperature you want to reach. To calculate the hydrocooling time for a specific type of produce, use this equation for DTD:

$$DTD = (T - W) \div (P - W)$$

where:

T = the target temperature (°F)

W = the temperature of the water (°F)

P = the starting temperature of the produce (°F)

Follow the steps in the example below to calculate the hydrocooling time for sweet corn.

Example: Sweet corn, with a center cob temperature of 85° F, is to be hydrocooled by immersion in 35°F water. How long will it take to reduce the center cob temperature to 55° F?

Figure must first be consulted to locate the cooling line (E) that applies to sweet corn. DTD is then calculated by using the above equation:

$$DTD = (55 - 35) \div (85 - 35) = 0.4$$

By finding the place on figure where curve E intersects the DTD = 0.4 line and projecting downward to the X-axis, you will see that the time required to cool the sweet corn from 85° F to 55° F is approximately 28 minutes. Suppose the packer was considering cooling the corn not to 55° F but to 42° F. In this case the DTD would be:

$$DTD = (42 - 35) \div (85 - 35) = 0.14$$

Again, by finding the place on figure where curve E intersects the DTD = 0.14 line and projecting downward to the X-axis, the time required to cool the sweet corn from 85° F to 42° F is found to be approximately 56 minutes.

In the example above, it was assumed that the individual ears of corn were completely immersed in agitated, chilled water. Recall that immersion is the fastest possible hydrocooling method. When the sweet corn is totally immersed, all outside surfaces are covered with cold water. If the sweet corn were hydrocooled in a conventional hydrocooler or closely packed in a wire-bound crate with other ears of corn, the cooling time would be considerably longer. Data published by the manufacturers of conventional hydrocoolers suggest that cooling times for their equipment be estimated at 40 to 80 percent longer than for immersion hydrocooling. The cooling time for sweet corn on a conventional hydrocooler can be estimated by multiplying the immersion hydrocooler rate by a conversion factor of 1.5:28 minutes (immersion hydrocooling rate) x 1.5 (conversion factor) = 42 minutes (conventional hydrocooling rate).

Figure should be used as a rough guide only. Consult a hydrocooler manufacturer for more precise hydrocooling rates.

Figure of cooling rate for sweet corn shows the cooling curve for sweet corn with a starting temperature of 85° F and a cooling water temperature of 35° F. Note that the cooling proceeds rapidly at first but then slows as the produce temperature approaches that of the chilled water. It takes 28 minutes to cool the corn 30 degrees (from 85° F to 55° F), but it takes twice that long to cool it another 13 degrees (from 55° F to 42° F). The graph illustrates that hydrocooling is much more efficient at removing the first 30 degrees or so of heat from the produce. Attempts to reduce the heat further by hydrocooling may result in decreased productivity and increased cooling costs. Alternate cooling methods, such as top icing, may be more practical if additional cooling is desired.

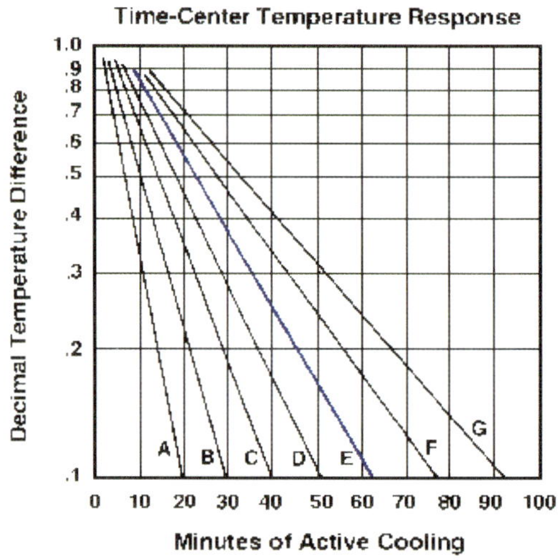

Figure: Time-temperature response of various fruits and vegetables immersed in agitated chilled water. Refer to the box for the KEY

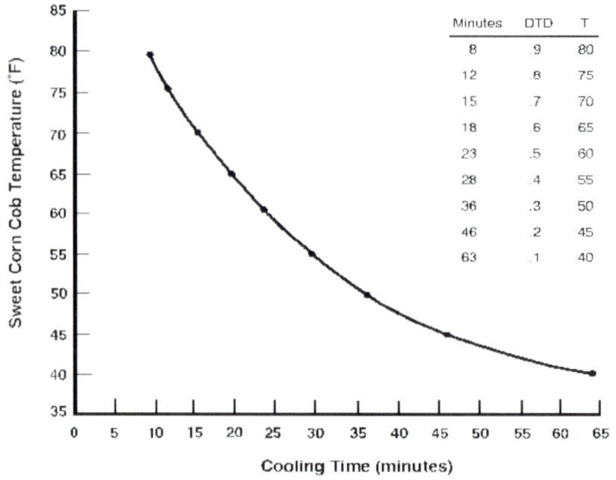

Figure: Example of cooling rate for sweet corn

Chlorination, Disease Control, and Wastewater Quality

Although hydrocooling is an excellent cooling method, it does wet the produce. The surface of warm, wet produce provides an excellent site for postharvest diseases. Therefore, it is essential that once hydrocooled, produce should not be allowed to rewarm. Produce is particularly susceptible to postharvest diseases when it is stressed by too much or too little water, high rates of nitrogen, or mechanical injury (scrapes, bruises, or abrasions). Also, since the hydrocooling water is recirculated, it can spread disease from a few infected items to all the produce hydrocooled thereafter. To reduce the spread of postharvest diseases, the *North Carolina Agricultural Chemicals Manual* recommends the use of chlorine as a disinfectant at the rate of 55 to 70 parts per million. Chlorine for use in hydrocooling water can be purchased in the form of a sodium hypochlorite solution (for example, Clorox) or as dry, powdered calcium hypochlorite.

Several factors affect the amount of chlorine available in the hydrocooling water over time. Chlorine used in hydrocooling water is quite volatile and will disperse into the air at a rate that increases with the temperature of the water. The warmer the water, the faster the chlorine will leave the solution. Furthermore, chlorine tends to attach to soil particles. Thus, dirty produce uses up the available chlorine much faster than relatively clean produce.

In addition to the factors discussed above, the pH of the water has a significant effect on the availability of chlorine. Even if a normally sufficient amount of chlorine is added to the hydrocooling water, the chlorine may not be available in a usable form if that water is acid (below a pH of 6.5) or basic (above a pH of 7.5). For the chlorine to be optimally available, the pH of the hydrocooling water should be nearly neutral (pH of 7.0). Over seven times more chlorine is needed to disinfect at a pH of 8.5 than at a pH of 7.0.

Because the pH of well water in North Carolina varies from moderately acid to moderately basic, it is a good idea to check the water with a pH meter or test papers. Even if your water has a nearly neutral pH, the addition of hypochlorites will cause the water to become basic. Therefore, adding a small amount of common acids like lemon juice or vinegar may be necessary to correct the pH. Inexpensive test papers for checking both the chlorine level and pH may be obtained from most swimming pool and chemical supply houses.

Pathogenic (disease-causing) organisms may enter the hydrocooling water both in the active vegetative form and in the form of spores. The chlorine will quickly kill the vegetative form, but the spores are 10 to 1,000 times more difficult to kill. Therefore, chlorine treatment does not usually eliminate all pathogens and sterilize the surfaces of the produce. Many spores may remain on the surface to develop later if the opportunity arises (that is, if the produce is allowed to rewarm). The effectiveness of the chlorine treatment depends on the length of exposure. Fortunately, the long exposure that is common with hydrocooling is much more effective than a quick dip treatment. However, chlorination is only a surface treatment. If the pathogens have already started to develop below the surface, chlorine will be ineffective. Also, chlorine solutions can produce surface bleaching.

The wastewater from hydrocooling is usually dumped at the end of each workday or more often if necessary. This wastewater often contains high concentrations of sediment, pesticides, and other suspended matter. Hydrocooler water may be considered an industrial wastewater if the product is discharged to a municipal wastewater treatment plant or to surface waters (canals, creeks, or ponds). Land application of this material is normally permitted, but a nondischarge permit may be required. A hydrocooler owner may be required to obtain a wastewater discharge permit.

Energy Efficiency

Well-designed and properly operated hydrocoolers have the advantage of being able to cool large amounts of produce in a short period of time. However, many commercial and most homemade hydrocoolers are not very energy efficient. Studies have shown that conventional and batch hydrocoolers can waste as much as one half the total energy input. Truck hydrocoolers can be even less energy efficient and may waste as much as 85 percent of the energy input. This is because water contacts the truck body and surrounding areas, substantially increasing the amount of heat entering the water.

During hydrocooling, much of the energy loss occurs because the falling water gains heat from the air. Energy loss also results from the lack of insulation on refrigerated surfaces. Additional losses occur if a hydrocooler is operated at less than full capacity, if it is operated intermittently, or if more water than necessary is used.

Immersion hydrocoolers are the most energy-efficient type of hydrocooler because the water-to-air contact is minimized. A well-designed immersion hydrocooler may have an energy efficiency as great as 70 percent.

Energy use in hydrocoolers may be substantially reduced by:

- Using generous amounts of insulation on all refrigerated surfaces and positioning the hydrocooler out of the wind and sunlight.
- Using plastic strips on both the inlet and outlet ends of conveyor hydrocoolers to reduce the water-to-air heat gain.
- Operating the hydrocooler at maximum capacity. Intermittent operation wastes energy, as does operation at reduced capacity. A low-capacity hydrocooler operated for long periods is more energy efficient and economical than a high-capacity unit operated for shorter periods. Some hydrocoolers may now be purchased with a thermal storage option. Thermal storage capacity not only reduces the size of the refrigeration unit necessary but may also result in a smaller energy bill.
- Using an appropriate-size water reservoir. An oversized water reservoir wastes energy because the remaining cooling capacity of the water is lost when it is dumped at the end of the day.

Vacuum Cooling

The process of vacuum cooling is the rapid refrigeration of a porous product through moisture removal. Once a product is cut, heat and moisture begin to take their toll. The product will begin to rot from the inside out. When placed in a cold room, the heat is removed in a slow continuous process. However with vacuum cooling the heat is removed very rapidly, allowing for earlier shipping and a fresher product.

After the product is harvested and packaged, it is placed on pallets for shipping. The pallets are loaded into a vacuum tight chamber and the process begins. The equipment drops the pressure inside the vessel by removing air through vacuum pumps. Once the level of vacuum reaches the "flash point" moisture and heat are removed from the product in the form of water vapor. The lower the pressure inside the chamber, the lower the boiling point of water. As the vacuum goes deeper, more water vapor is removed and the product temperature is reduced until the desired temperature is achieved.

The water vapor must be removed in order for the process to work efficiently. The vacuum draws the water vapor past refrigeration coils which condenses the vapor back into water. These two systems allow the process to achieve desired product temperature more rapidly than any other post harvest cooling technique.

The rate at which a product can be refrigerated is directly related to the surface area, density of its tissue and the amount of temperature drop. Typical cooling times range from 20 to 40 minutes at a temperature drop from 80 to 36 degrees Fahrenheit. The average moisture loss is one-percent for each eleven-degree Fahrenheit temperature drop.

Since produce is sold by weight, a hydro-vac system can help reduce moisture loss. This works by spraying clean water over the produce during the vacuum cooling cycle.

Due to the rapid post harvest refrigeration that vacuum cooling allows, shelf life of the product is extended considerably, thus increasing allowable travel times and opening new markets. Rapidly cooling products will greatly reduce the level of microbiological growth during storage. Vacuum cooling has been documented to have lower levels of microbiological growth than any other cooling method.

Frost, Freezing and Chilling Injury

The lower desert areas experience freezing temperatures several times during the winter and it is not uncommon winter vegetables injured. There are two types of injury that low temperatures can cause on vegetables. Chilling injury is what happens to some vegetable crops of tropical origin held at the wrong storage or transit temperature, but a temperature above 32° F (0° C). Generally these are temperatures around 41-50° F (12.5° C). Chilling injury occurs at temperatures well above freezing point. The tissue becomes weakened leading to cellular dysfunctions. Symptoms include surface lesions/pitting, internal discoloration, water soaking of the tissue, failure to ripen normally and increased susceptibility to decay organisms such as Alternaria. Maturity at harvest also affects the susceptibility to chilling injury in products such as tomatoes, honeydew melons and peppers.

The second type of injury is frost/freezing injury and it can occur in a field when temperatures drop to or below 32° F (0° C). It can also occur during cold storage if temperatures were below the freezing point of the product. This exposure to freezing temperatures may have a drastic effect upon the entire plant or affect only a small part of the plant tissue, resulting in reduced yields or poor product quality. Ice formation on the plant and in the plant tissue is what causes the damage, especially when the tissue thaws. Postharvest researchers have worked on identifying the freezing temperatures for most crops and have published tables with the recommended optimum storage temperatures. No vegetable or fruit cold storage facility should be without these tables and they are available for download at the UC Davis Postharvest web site. Researchers have categorized fruits and vegetables into three groups based on their sensitivity to freezing: most susceptible are those that are likely to be injured by one light freeze, moderately susceptible are those that will recover from one or two light freezes and least susceptible are those that can be lightly frozen several times without serious injury.

Most susceptible,

- Asparagus
- Beans, snap
- Cucumbers
- Eggplant
- Lemons
- Tomatoes
- Lettuce
- Limes
- Okra
- Peppers, sweet
- Potato

Moderately susceptible,

- Broccoli
- Carrots
- Cauliflower
- Celery
- Grapefruit
- Grapes
- Onion (dry)
- Oranges
- Parsley
- Radishes
- Spinach
- Squash

Least susceptible,

- Brussels sprouts
- Cabbage, mature and savory

- Dates
- Kale
- Kohlrabi
- Parsnips
- Turnips
- Beets

Symptoms of freezing, chilling injury

Artichoke: Freezing injury will be initiated at 29.9° F (-1.2° C). Symptoms of light freezing injury are blistering of the cuticle and a bronzing of the outer bracts. This may occur in the field with winter-harvested buds. More severe freeze injury results in water soaked bracts and the heart becoming brown to black then gelatinous in appearance over time.

Asparagus: Freezing injury (water-soaked appearance leading to extreme softening) will likely result at temperatures of 30.9° F (-0.6° C) or lower.

The tip becomes limp and dark; the rest of the spear is water-soaked. Thawed spears become mushy. Chilling Injury occurs when spears are held more than 10 days at 32° F (0° C) and symptoms of chilling injury include loss of sheen or glossiness and graying of the tips. A limp, wilted appearance may be observed. Severe chilling injury may result in darkening near tips in spots or streaks

Snap Bean: Freezing Injury appears as water-soaked areas that subsequently deteriorate and decay. Freezing injury occurs at temperatures of 30.7° F (-0.7° C) or below. The typical symptom of chilling injury in snap beans stored below 41° F (5° C) but above freezing point for longer than 5-6 days is a general opaque discoloration of the entire bean. A less common symptom is pitting on the surface. The most common symptom of chilling injury is the appearance of discrete rusty brown spots which occur in the temperature range of 41-45° F (5-7.5° C). These lesions are very susceptible to attack by common fungal pathogens. Beans can be held about 2 days at 34° F (1° C), 4 days at 36° F (2.5° C), or 8-10 days at 41° F (5° C) before chilling symptoms appear. No discoloration occurs on beans stored at 50° F (10° C). Different snap bean varieties differ significantly in their susceptibility to chilling injury.

Bell Pepper: Freezing injury symptoms include dead, water-soaked tissue in part or all of the pericarp surface; pitting, shriveling, and decay follow thawing. Symptoms of chilling injury include surface pitting, water-soaked areas, decay (especially Alternaria spp.), and discoloration of the seed cavity.

Broccoli: will freeze if stored at 30.6° F (-0.6° C) to 30° F (-1.0°C). This may also occur if salt is used in the liquid-ice cooling slurry. Frozen and thawed areas on the florets appear very dark and translucent, may discolor after thawing and are very susceptible to bacterial decay. The youngest florets in the center of the curd are most sensitive to freezing injury. They turn brown and give off strong odors upon thawing.

Cabbage: Freeze damage appears as darkened translucent or water-soaked areas that will deteriorate rapidly after thawing. Freeze damage can occur if round cabbages are stored below 30.4°

F (-0.9° C) and if Chinese cabbage is stored below 31° F (-0.6° C). Leaves become water-soaked, translucent, and limp upon thawing; the epidermis can also separate from the leaf as it does in lettuce.

Carrot: Freeze damage includes blistered appearance, jagged length-wise cracks. Carrot interior becomes water-soaked and darkened upon thawing. Freeze damage occur at below 29.5° F (-1.4° C) but can vary depending on the sugar content of the carrot.

Cauliflower: Freezing injury will be initiated at 30.6° F (- 0.8° C). Symptoms of freezing injury include a water-soaked and greyish curd and water-soaked or wilted crown leaves. The curd will become brown and gelatinous in appearance following invasion by soft-rot bacteria. These brown curds have a strong off-odor when cooked.

Celery: Freezing injury will be initiated at 31.1° F (- 0.5° C). Symptoms of freezing injury include a water-soaked appearance on thawing and wilted leaves. Mild freezing causes pitting or short streaks in the petiole which develop a brown discoloration with additional storage. Leaves and petioles appear wilted and water-soaked upon thawing. Petioles freeze more readily than leaves.

Eggplant: Freezing injury will be initiated at 30.6° F (-0.8° C), depending on the soluble solids content. Symptoms of freezing injury include a water-soaked pulp becoming brown and desiccated in appearance over time. Eggplant fruit are chilling sensitive at temperatures below 50° F (10° C). At 41° F (5° C) chilling injury will occur in 6-8 days. Consequences of chilling injury are pitting, surface bronzing, and browning of seeds and pulp tissue. Accelerated decay by Alternaria spp. is common in chilling stressed fruit. Chilling injury is cumulative and may be initiated in the field prior to harvest.

Beet: Freezing injury symptoms include external and internal water-soaking; sometimes blackening of conducting tissue.

Garlic: Freezes at temperatures below 28° F (-2° C) due to its high solids content. Thawed cloves appear grayish-yellow and water-soaked.

Green Onion: Freezing injury will be initiated at 30.6° F (-1° C). Symptoms of freezing injury include a water-soaked appearance of bulb or leaves and wilted or gelatinous leaves, after thawing. The bulb will become soft or gelatinous in texture in outer tissue. Freeze injury is rapidly followed by bacterial soft-rot decay.

Lettuce Romaine and Crisphead: Freeze damage can occur in the field and cause separation of the epidermis from the leaf. This weakens the leaf and leads to bacterial decay during storage. Freeze damage can occur during storage if the lettuce is held at 31.7° F (<-0.2° C). This appears as darkened translucent or water-soaked areas that will turn slimy and deteriorate rapidly after thawing. The blistered dead cells of the separated epidermis on outer leaves become tan and there is increased susceptibility to physical damage and decay.

Okra: Freeze damage occurs at temperatures of 28.7° F (-1.8° C) or below. The typical symptoms of chilling injury in okra are discoloration, pitting, water-soaked lesions and increased decay (especially after removal to warmer temperatures, as during marketing). Different cultivars may differ in their susceptibility to chilling injury. Calcium dips and modified atmospheres have been reported to reduce chilling symptoms.

Onion: Freezing injury symptoms include soft water-soaked scales that rapidly decay from subsequent microbial growth. Thawed bulbs are soft, greyish-yellow, and water-soaked in cross section; often limited to individual scales.

Potato: Freezing injury will be initiated at 30.5° F (-0.8° C). Symptoms of freezing injury include a water-soaked appearance, glassiness, and tissue breakdown on thawing. Mild freezing may also result in chilling injury. Freezing injury may not be externally evident, but shows as grey or bluish-grey patches beneath the skin. Thawed tubers become soft and watery. Chilling injury can occur at storage temperatures near 32° F (0° C) after a few weeks and may result in a mahogany discoloration of internal tissue and eventually complete internal breakdown. Much longer periods (months) of storage are generally required for chilling injury at higher temperatures 36-41° F (2-5° C).

Radish: Radish ideally stored and transported just above the freezing point 30.5° F (-1.0° C), but freeze injury is not uncommon. Shoots become water-soaked, wilted, and turn black. Roots appear water-soaked and glassy, often only at the outer layers if the freezing temperature is not too low. Roots become soft quickly on warming and pigmented roots may "bleed" (lose pigment). Thawed tissues appear translucent; roots soften and shrivel.

Spinach: Freezing injury will be initiated at 31.5° F (-0.3° C). Freezing injury results in water soaking typically followed by rapid decay by soft-rot bacteria.

Tomato: Freezing injury will be initiated at 30° F (-1° C), depending on the soluble solids content. Symptoms of freezing injury include a watersoaked appearance, excessive softening, and desiccated appearance of the locular gel. In partially frozen fruits, the margin between healthy and dead tissue is distinct, especially in green fruits. Tomatoes are chilling sensitive at temperatures below 50° F (10° C) if held for longer than 2 weeks or at 41° F (5° C) for longer than 6-8 days. Consequences of chilling injury are failure to ripen and develop full color and flavor, irregular (blotchy) color development, premature softening, surface pitting, browning of seeds, and increased decay (especially Black mold caused by Alternaria spp.). Chilling injury is cumulative and may be initiated in the field prior to harvest.

Turnip: Freezing injury symptoms include small water-soaked spots or pitting on the surface. Injured tissues appear tan or gray and give off an objectionable odor.

Minimizing Physical Damage

By autumn, most summer vegetable crops are in their peak of maturity and it's critical to harvest the crop at the right stage to ensure high quality.

When harvesting vegetables, avoid any physical damage to the fruit skin as it minimizes the storage quality and could lead to decay issues. If the fruits are tough to harvest by hand, use a sharp knife to cut the fruit stalk. This minimizes any accidental tearing of the stem ends. It is best to harvest the crops when the weather is dry, preferably during the later part of the day after the morning dew has dried.

Tomatoes: For juice or canning purposes, harvest tomatoes that are fully ripe and colored. Mature fruits that have growth cracks on the top should be harvested when they turn partial pink in color

(30-60 percent) and stored indoors at normal room temperature for about six days to obtain their full color. Before frost, harvest any mature green tomatoes and store them indoors for ripening.

Sweet potatoes: Harvest sweet potatoes just before the first frost. Beginning in late August, allow the vine to stress out by not feeding with any fertilizer or watering the vine. This helps in translocating the starch from the vine to the tubers. Use a digging fork to carefully uproot the tubers, and try to minimize any bruises on the tubers. Cure the harvested tubers indoors at normal room temperature for about two to three weeks.

Pumpkins and winter squash: Pumpkins and winter squash begin to mature from September through October. Harvest the fruit at their full maturity when the fruit rind becomes hard and waxy with a uniform deep solid color. To test the firmness of the rind, gently apply fingernail pressure and check the resistance of the fruit.

Peppers: Green bell peppers are harvested when the mature green fruits are fully grown to about 3-4 inches long, firm and tend to break easily from the plant. Depending on the weather conditions in autumn, colored bell peppers can take a while to ripen (develops red, orange, or yellow fruit color at its ripe stage). Hot peppers are harvested when the fruits are reddish in color.

Sweet corn: Harvest the kernels about 21 days after silky strands appear. The kernels should be fully formed and milky. Other harvest indicators are browning of the silks and kernels are fully formed at its tip.

References

- Frantzen, Allen J. (2014). Food, Eating and Identity in Early Medieval England. Boydell & Brewer Ltd. p. 65. ISBN 978-1-84383-908-8
- Storage-temperature-fruits-vegetables: washingtoncoldstorage.com, Retrieved 11 May 2018
- England, Historic. "Historic Environment Local Management Training Programme - Historic England". www.helm.org.uk. Archived from the original on 9 August 2012. Retrieved 5 May 2018
- The-effects-of-temperature-and-moisture-on-stored-grains, food-storage: preparednessadvice.com, Retrieved 23 June 2018
- Malcolm Kirk, The Barn. Silent Spaces, London 1994; Graham Hughes, Barns of Rural Britain, London 1985; Walter Horn, 'On the Origins of the Medieval Bay System', in: Journal of the Society of Architectural Historians 17 (1958), nr. 2, p. 2-23
- Symptoms-frost-freezing-and-chilling-injury-vegetables: westernfarmpress.com, Retrieved 13 July 2018
- Harvest-vegetable-crops-right-time-best-quality-625288001: greenbaypressgazette.com, Retrieved 23 April 2018

4

Postharvest Packaging

Science and technology have undergone rapid developments in the past decade, which has resulted in improved techniques of postharvest packaging in the field of agriculture. This chapter extensively discusses about the fundamentals of food packaging, use of sanitizers and disinfectants, removal of microorganisms, artificial waxing, preservation, vacuum packaging, food fortification, etc.

Food Packaging

Food packaging is defined as a co-ordinated system of preparing food for transport, distribution, storage, retailing, and end-use to satisfy the ultimate consumer with optimal cost. Food packaging is an essential part of modern society; commercially processed food could not be handled and distributed safely and efficiently without packaging. The World Packaging Organization (WPO) estimates that more than 25% of food is wasted because of poor packaging. Thus, it is clear that optimal packaging can reduce the large amount of food waste. Moreover, the current consumer demand for convenient and high-quality food products has increased the impact of food packaging.

Functions of Food Packaging

Containment

The term "containment" means, simply, to contain products to enable them to be moved or stored. It is so basic that it is easily overlooked. However, containment is a key factor for all other packaging functions. All products must be contained for delivery from their point of production to their ultimate destination. Even items that consumers consider as "not a packaged product," such as bulk produce, must be packaged for transportation. Without packaging, products are likely to be lost or contaminated by the environment. For this reason, we have actually used packaging for millennia. Early packaging such as animal skins, baskets, or leaves from trees were used to contain liquids, powders, grains, etc. The containment function significantly contributes to protecting and preserving products during their distribution.

Protection/Preservation

There are two broad types of damage that fresh and processed foods sustain during storage and transportation. One is physical damage such as shock, vibration, compressive forces, etc. The other is environmental damage that occurs due to exposure to water, light, gases, odors, microorganisms, etc. A good packaging system will protect or reduce these types of damage to the package contents. For example, an essential aroma or flavor in coffee or juice may easily be evaporated or oxidized without optimum barrier packaging. A shelf-stable food in a can or pouch may maintain its stability (especially against microorganisms) as long as the package provides protection.

However, in the case of fresh food products, the ideal protection is usually hard to achieve with packaging alone. Since temperature is a major influence on the degradation of food, it is more economical to control temperature through supply chain modification (refrigeration, freezing, etc.). However, packaging can also add a certain level of protection to slow down temperature changes.

Communication

According to the Fair Packaging and Labeling Act, food packaging must identify the product, the net quantity of the contents, name/address of business of the manufacturer, packer, or distributor, as well as (usually) nutritional information. The communication function of packaging not only includes the information provided by the written text, but also elements of the packaging design such as package shape, color, recognized symbols or brands. Beyond giving information, the communication function is expected to entice the consumer to purchase the product. Packaging has been regarded as the "silent salesman". Consumers may instantly recognize products through appetizing pictures or distinctive brands on packaging, and even simple transparency of the packaging material can attract consumers by allowing them to view the product inside.

Another aspect of the communication function is also important. The Universal Product Code (UPC) is widely used to facilitate rapid and accurate checkout in retail stores. Also, most warehouse and distribution centers track and manage their inventory using UPCs. Currently, by using radiofrequency identification (RFID) tags attached to secondary and tertiary packages, manufacturers are able to get better demand signals from customers and markets. An RFID tag can gather data on items automatically without human intervention or data entry. It identifies, categorizes, and manages product and information flow at important inspection and decision points. In addition, RFID tags can be read all at once (e.g. up to 50 per second), while UPC codes can only be read one at a time. RFID technology looks promising in terms of revolutionizing the supply chain. However, it needs to be economical. The cost of individual tags and antenna reading systems is still high. Another problem is the lack of uniformity in global standards in the area of sensor technology. Sensor providers usually provide their own interfaces to communicate with their own tags.

Utility

This function of packaging is sometimes termed "convenience." Consumers demand products that fit into their lifestyles and the packaging industry has had to respond to this. Thus, the utility function encompasses all the packaging attributes that provide added value and convenience to the users of the product and package. For example, an important social trend is the growing number of mothers in the workforce and smaller households (people living alone and married couples without children). Unquestionably, food products that offer simplification and convenience have grown in popularity with this group; examples include microwavable entrees, steam-in-pouch vegetables, oven-safe meat pouches, pump-action condiments, and so on.

Packaging Systems

We can categorize packaging systems into four groups: primary packaging, secondary packaging, distribution or tertiary packaging, and unit load.

Primary Packaging

The first-level package that directly contacts the product is referred to as the "primary package." For example, a beverage can or a jar, a paper envelope for a tea bag, an inner bag in a cereal box, and an individual candy wrap in a pouch are primary packages, and their main function is to contain and preserve the product. Primary packages must be non-toxic and compatible with the food and should not cause any changes in food attributes such as color changes, undesired chemical reactions, flavour etc.

Secondary Packaging

The secondary package contains two or more primary packages and protects the primary packages from damage during shipment and storage. Secondary packages are also used to prevent dirt and contaminants from soiling the primary packages; they also unitize groups of primary packages. A shrink wrap and a plastic ring connector that bundles two or more cans together to enhance ease of handling are examples of secondary packages.

Tertiary Package

The tertiary package is the shipping container, which typically contains a number of the primary or secondary packages. It is also referred to as the "distribution package."

A corrugated box is by far the most common form of tertiary package. Its main function is to protect the product during distribution and to provide for efficient handling.

Unit Load

A unit load means a group of tertiary packages assembled into a single unit. If the corrugated boxes are placed on a pallet and stretch wrapped for mechanical handling, shipping and storage, the single unit is referred to as a "unit load." The objective is to aid in the automated handling of larger amounts of product. A fork-lift truck or similar equipment is used to transport the unit load.

Consumer/Industrial Packaging

Packaging systems can also be categorized into consumer and industrial packaging. Consumer packaging means a package that will be delivered to the ultimate consumer in the retail store. Usually, primary and secondary packages fit in this category. Industrial packaging means a package for warehousing and distribution to the retail store. Tertiary packages and unit loads fit in this category.

Not all package systems are actually composed of a set of primary, secondary, and tertiary packages. For example, the packaging system for potato chips usually consists only of a flexible barrier bag and a corrugated shipping container before they are palletized, while mayonnaise jars are sold in a club store as a two-pack consisting of plastic bottles, shrink wrap, corrugated boxes, and pallet. Often, the distinction between consumer and industrial packaging is more clear-cut than between primary, secondary, and tertiary packaging.

Aseptic Packaging

Aseptic packaging is the filling of a commercially sterilized product into a sterilized container under aseptic conditions and then sealing it hermetically to prevent contamination. Therefore, the sterility can be maintained throughout the handling and distribution process. The aseptic packaging permits shorter heat exposure for the food than typical thermal sterilization processes such as canning and retorting; therefore it generally results in superior food quality compared to typical thermal sterilization. Because of the separate package and food product sterilization, the selection of material and container design can be more flexible than in traditional thermal sterilization.

The required microbial count reduction for the sterilization of a food contact packaging material is determined by the type of product. For nonsterile acidic products (pH <4.5), a minimum 4-log reduction in bacterial spores is required. For sterile, neutral, low-acid products (pH >4.5), a 6-log reduction is required. In addition, if sterility against Clostridium botulinum endospores is required, a 12-log reduction is required, and the process is called commercial sterilization. The sterilization methods for aseptic packaging materials include irradiation (such as ultraviolet rays, infrared rays, and ionizing radiation), heat (such as saturated steam, superheated steam, hot air, hot air and steam), and chemical treatment (such as hydrogen peroxide, ethylene oxide, peracetic acid). These processes can be used either individually or in combination.

The type of aseptic packaging material used is influenced by the nature of the product, the cost of both product and packaging, and the preference of consumers. The most widely used aseptic package is the paperboard laminated carton. The typical structure of this material consists of unbleached or bleached paperboard, polyethylene, and aluminum foil. The laminated structure is impermeable to liquid, gas, and light.

Aseptic packages can also be in the form of cans, bottles, pouches, trays, or cups. Can type aseptic packages are the same basic types of metal cans as in the regular canning process: tinplate, ECCS, and aluminum. Under an aseptic packaging process, the heating time for sterilization and food quality are the same for both large and small containers, while the traditional canning process has longer heating times for larger containers. Bottle type aseptic packages are produced from plastics as an economical alternative to glass for non-returnable containers. HDPE, PP, and PET are the most commonly used materials for this type of package. Pouch type aseptic packages are made from similar laminated paperboard structures. Also, multilayer films containing LLDPE and EVOH are used. Cup type aseptic packages are produced from HIPS, PP, or multilayer film. Multilayer films are the choice when high barrier properties are required. PVDC or EVOH is typically used as the barrier layer.

Maintaining the package integrity during distribution and handling is one of the most important issues in aseptic food packaging. Thus, various integrity tests are used commercially. Typically, electrolytic testing, dye penetration or vacuum leak tests are used for quality control during manufacturing. However, these traditional methods are destructive and therefore it is impossible to test and reject all faulty packages in the processing line. Thus, non-destructive test methods, such as gas leak detectors and ultrasound techniques, are gaining more attention from industry.

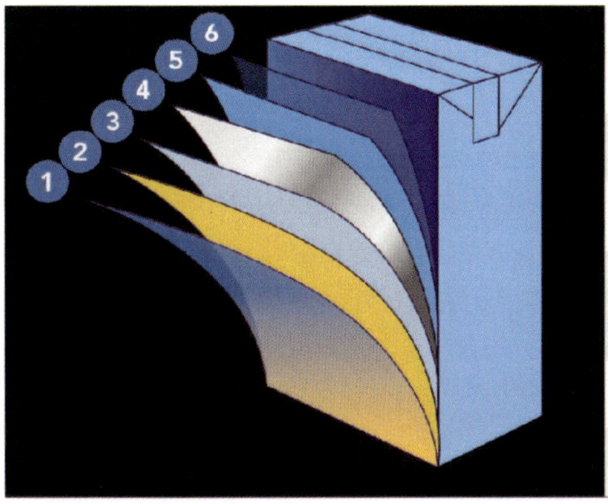

1 Polyethylene - protects against outside moisture

2 Paper - for stability and strength

3 Polyethylene - adhesion layer

4 Aluminium foil - oxygen, flavour and light barrier

5 Polyethylene - adhesion layer

6 Polyethylene - seals in the liquid

Figure: The structure of a typical laminated paperboard carton for aseptic packaging.

Modified Atmosphere Packaging

Modified atmosphere packaging (MAP) is based on modifying or altering the atmosphere inside the package to prolong shelf life and maintain quality of products. The modification of the atmosphere can be achieved actively or passively. In active type MAP, the optimum gaseous environment is obtained by flushing a controlled mixture of gases in a package ("gas flush"). The passive type of MAP modifies the optimum gaseous environment in a package by a combination of the food's respiration and the metabolism of microorganisms associated with the food and the permeability of the packaging. With the optimized gaseous atmosphere, degradation reactions in foods such as enzyme activity, oxidation, moisture loss, and postharvest metabolic activities as well as the growth of microorganisms are delayed. The three main gases used for MAP are nitrogen (N_2), carbon dioxide (CO_2), and oxygen (O_2). They are used either alone or, in most cases, in combination. Mixtures of carbon dioxide (CO) and argon (Ar) are also utilized commercially. Examples of gas mixtures that are used for fresh and processed foods are shown in Table.

Table: Examples of optimum headspace gas mixtures (%) and storage temperature for fresh and processed food products.

Product	Temperature (^{O}C)	Oxygen (%)	Carbon dioxide (%)	Nitrogen (%)
Snack	23	0	20–30	70–80
Bread	23	0	60–70	30–40
Cake	23	0	60	40
Cheese (hard)	4	0	60	40
Cheese (soft)	4	0	30	70
Pasta	4	0	80	20
Pizza	5	0–10	40–60	40–60
White fish	0–2	30	40	30
Oily fish	0–2	0	60	40
Shrimp	0–2	30	40	30
Fresh red meat	0–2	40–80	20	0
Cooked/cured meat	1–3	0	30	70
Pork	0–2	40–80	20	0

Poultry	0–2	0	20–100	
Sausage	4	0	80	20
Apples	0–3	1–3	0–3	
Banana	12–15	2–5	2–5	
Broccoli	0–5	5–10	1–2	
Lettuce	0–5	2–3	5–6	
Tomato	7–12	4	4	

Nitrogen (N_2)

Nitrogen is the most commonly used gas in MAP. It is an inert gas with no odor or taste. Nitrogen also has low solubility in water (0.009 g/kg at 20° C). Nitrogen does not directly provide any microbial retardation but it delays aerobic microbial growth and oxidation by replacing oxygen.

Carbon dioxide (CO_2)

Carbon dioxide is colorless with a slightly pungent odor. The most important function of this gas is related to its bacteriostatic and fungistatic properties. Usually, a higher concentration is more effective against microorganisms. However, color changes and acid tastes have been reported by several researchers due to exposure to high CO_2. Carbon dioxide has high solubility in water (1.69 g/kg at 20° C). Thus, packages containing moist foods with CO_2 in their headspace may collapse. 11.5.2.0.3 Oxygen (O_2) Oxygen is a colorless, odorless, and highly reactive gas. It promotes food deteriorative reactions such as fat oxidation, browning reactions, and aerobic microbial growth. For this reason, the desired oxygen content in the headspace of most foods is often extremely low. However, oxygen is still needed for the retention of color in red meat and for fruit/vegetable respiration.

Carbon Monoxide (CO)

Carbon monoxide is a colorless, tasteless, and odourless gas. When carbon monoxide combines with myoglobin, a bright red pigment (carboxymyoglobin) is formed, which is more stable than normal red meat pigment (oxymyoglobin). The use of carbon monoxide is not for quality issues but for its visual effect. Since carbon monoxide is a toxic and highly flammable gas, its commercial use is limited. In the US, carbon monoxide is currently used for red meat but the level of its use is limited to 0.4%.

Argon (Ar)

Argon is a chemically inert, colorless, tasteless, and odourless gas which is denser and heavier than air. Compared to nitrogen, argon is a more effective gas for use in flushing out air. However, due to the high cost of argon, nitrogen is mostly used for flushing.

Modified Atmosphere Packaging for Meat

For red meat, oxygen is necessary to maintain the red color of oxymyoglobin in unprocessed meats. On the other hand, a low oxygen level is desired to prevent the growth of microorganisms and oxidative rancidity of fat. Typically, the shelf life of fresh red meat is extended by packaging it in an atmosphere of 20% CO_2, 60–80% O_2, and up to 20% N2. However, off-odors and rancidity have been reported in meats stored at high O_2 concentrations. Thus, maintaining a low temperature (0–2° C) is desired when high-O2 MAP is used. Poultry has low myoglobin content so it does not need oxygen to maintain its color. A higher CO_2 concentration (20–100%) is possible to extend the shelf life. However, a few studies report color change with high CO_2 concentration in poultry products.

Modified Atmosphere Packaging

For seafood Seafood such as fish and shellfish is highly perishable owing to its high water activity (aw), pH, and the presence of autolytic enzymes, which cause rapid development of undesirable odors. A low oxygen atmosphere can delay the occurrence of undesirable odor and growth of aerobic microorganisms. For example, packaging tilapia fillets in 75% CO2:25% N2 extended their shelf life up to 80 days at 4° C. However, fresh fish may be contaminated with the anaerobic C. botulinum either as a result of being present in the microbiota of the fish ecosystem or due to postcatch contamination during processing. A low oxygen atmosphere poses a potential threat for a packaged fish to become toxic prior to spoilage. To assure the safety of MAP fish products, the product must be maintained at or below 3° C at all times. Since chilled storage control is a critical factor to determine the shelf life, some companies use time temperature indicators (TTI) to monitor for temperature abuse.

Modified Atmosphere Packaging for Fresh

Fruits and vegetables Fresh fruits and vegetables keep consuming O_2 and emitting CO_2 even after harvest. The purpose of MAP for fresh fruits and vegetables is to minimize the respiration and senescence without causing suffocation and damage to metabolic activity that rapidly reduces their shelf life.

However, a low oxygen and high CO_2 atmosphere that develops inside the package due to respiration of the productmayresult in the accumulationof ethanolandacetaldehyde, and fermentation could start. Thus, the package material needs to be somewhat permeable to oxygen and CO_2 to allow the transfer of the gases from outside and inside. The change in gas composition during storage dependsonthe permeability of the container to water vapour and gas, storage temperature, and the mass of the food.

Active Packaging in Food Processing

Active packaging is an important and rapidly growing area. There are several different definitions that can be found in the literature for active packaging. This type of packaging usually involves an interaction between the packaging components and the food product beyond the inert passive barrier function of the packaging material. The major active packaging technologies include oxygen scavengers, moisture absorbers, antimicrobial agent releasers, ethylene scavengers, flavor/odor absorbers, and temperature control packaging. In order to apply this technology, the major deteriorative factor(s) for food products should be understood. For example, the shelf life of a packaged food is affected by numerous factors such as acidity (pH), water activity (aw), respiration rate, oxidation, microbial spoilage, temperature, etc. By carefully considering all of these factors, active packaging can be developed and applied to maintain the quality of the product and to extend its shelf life. Active agents are contained in sachets or incorporated directly into packaging containers.

Oxygen Scavengers

Oxygen scavengers are the most commercially applied technology in the active packaging market. By scavenging oxygen molecules in the package headspace, oxidative damage to food components such as oil, flavors, vitamins, color, etc. can be prevented. In addition, reduced O_2 concentration in the package retards the growth of aerobic bacteria and mold.

The basic principle of oxygen scavenging is related to oxidation of the scavenging agents (either metallic or non-metallic based materials) to consume oxygen. The most common O_2 scavengers used in the food industry are sachets with iron powder, which are highly permeable to oxygen. By using iron powder, the O_2 concentration in the headspace can be reduced to less than 0.01% while vacuum or gas flushing typically achieves 0.3–3.0% residual O_2 levels. The metal-based oxygen scavengers normally cannot pass the metal detectors on the packaging lines and cannot provide transparent packages if they are directly incorporated into packages. Non-metallic O_2 scavengers, such as ascorbic acid or glucose oxidase, can be used as an alternative choice. However, the use of non-metallic O_2 scavengers is not widespread.

O_2 scavengers can be incorporated into plastic film/sheet when there are market concerns about accidental consumption of sachets or when an O_2 scavenger needs to be used for liquid foods. In this case, the O_2 scavenger impregnated layer is typically sandwiched between film layers. The outside layer provides high oxygen protection and the inner layer prevents direct contact between the scavenger-containing matrix layer and the food.

Moisture Absorbers

Moisture in packages is a major cause of food deterioration such as microbial growth-related spoilage and product softening. Moisture-absorbing sachets are used in food packaging for humidity control. Several desiccants such as silica gel, calcium oxide, and activated clays and minerals are typically contained in sachets. Dripabsorbent pads and sheets are also used to absorb liquid in high aw foods such as meats, fish, poultry, fruits, and vegetables. A superabsorbent polymer is sandwiched by a microporous non-woven plastic film such as polyethylene or polypropylene. Polyacrylate salts, carboxymethyl cellulose (CMC), and starch co-polymers are typically used as the absorbent polymer.

Antimicrobial Agent Releasers

Typically, surface contamination during food handling and transportation is one of the most common sources of food-borne illness, and lower amounts of antimicrobial agents are required to control the surface microbial contamination/growth if they are incorporated into or onto packaging material rather than directly added into the food itself. There are two mechanisms for antimicrobial action: one is controlling microbial growth by slow and controlled release (or migration) of the antimicrobial agents over the product shelf life. The other type is controlling microbial growth by contact without release of the antimicrobial agents, which are immobilized on the surface of the package. Despite the large number of experimental studies on antimicrobial packaging, the technology has not been widely used in food markets yet, because of cost or regulatory constraints. An example of a commercial application of an antimicrobial packaging is silver ion-based film. This film is an effective antimicrobial agent with very low human toxicity and it has been used in food containers. Other examples of antimicrobials used (or studied) include nisin, pediocin, organic acids, grapefruit seed extract, cinnamon, and horseradish.

Ethylene Scavengers

Ethylene works as a plant growth regulator which accelerates the respiration rate and senescence of fruits and vegetables. Removing ethylene from the environment can extend the shelf life of horticultural products. One of the most common ethylene scavengers is made from potassium permanganate ($KMnO_4$) immobilized on an inert mineral substrate such as alumina or silica gel. Activated carbon-based scavengers with various metal catalysts are also used as effective ethylene scavengers. In recent years, packaging films and bags have been commercialized based on the reputed ability of certain finely dispersed minerals to absorb ethylene (such as clays, zeolites, coral, ceramics, etc.).

Flavor and Odor Absorbers

Undesirable flavor scalping by the packaging material can result in the loss of desirable food flavors, while foods can pick up undesirable odors or flavors from the package or the surrounding environment. For example, fruit juices in PET bottles or water in HDPE bottles can absorb unwanted odors that result from aldehydes such as hexanal and heptanal originating in oxidation of the plastics. Another example is that an unpleasant "confinement odor" can accumulate inside PE bags in a distribution center. One type of flavor/odor-absorbing technology is based on a molecular sieve with pore sizes of approximately 5 nm. Cyclodextrin is a common example of an odor absorber. Wood reported that cyclodextrin, grafted to a PE layer, was applied to military ration pouches to absorb accumulated odor from food decomposition. Synthetic aluminosilicate zeolite, which has a highly porous structure, has been incorporated in packaging materials, especially papers, to absorb odorous gases such as aldehyde.

On the other hand, absorbers can also be used in a flavor/ aroma-releasing system. Aroma compounds can be released inside the package to enhance the flavor/aroma of the product or outside the package to attract consumers to the products on a retail store shelf.

Temperature-controlled Packages

Temperature-controlled active packages generally include self-heating and self-cooling systems. The fundamental concepts for self-heating are not new. An exothermic chemical reaction between CaO (quicklime) and a waterbased solution generates heat. The challenging part of this system is optimizing the reaction and the thermal design of the container to provide an efficient, safe, and cost-effective package. Self-heating packages are commercially available for coffee, tea, and ready-to-eat meals. Self-cooling cans use endothermic chemical reactions; the latent heat of evaporating water is often used to produce the cooling effect. One example is dissolution of ammonium nitrate and chloride in water to cool the product.

Sustainable Food Packaging

Recycling of food packaging Food packaging has the largest demand for packaging industries, whether it is paper, plastics, glass, or metal. Finding or improving ways to reduce landfilled waste is important to meet current demands for environmentally friendly packaging. Recycling can be defined as diverting materials from the solid waste stream for use as raw materials in the manufacture of new products. The overall recycling rate of packaging material in the US is 40%, and is far behind the rate in Europe, which is about 59%.

Recycling of Paper and Paperboard

Packaging materials are the largest sector in which recycled paper is used in the US, as shown in figure. The recycling rate for paper and paperboard has been increasing during the last decade. In 2010, 37.7 million tonnes of paper and paperboard packaging waste were generated and 71.3% of the used material was recycled. Even if most types of paper are recyclable, recycled paper is not suitable for most food contact packaging applications because the recycling processes may allow contaminants to be present in the recycled paper products. In many applications, recycled paper is not used in direct contact with food, so migration is of little or no concern. Another issue is related to paper properties: recycled paper normally has a damaged and weak fiber structure. In order to obtain suitable physical and mechanical characteristics, mixing virgin and recovered fibers in different proportions is often required.

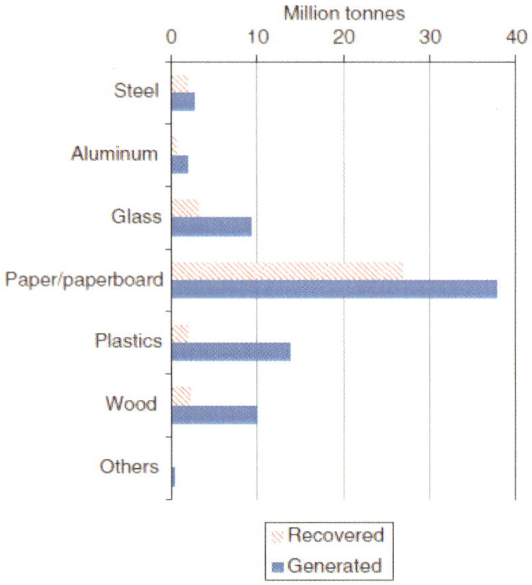

Figure: Generation and recovery of packaging materials in the United States.

Recycling of Glass

Glass is the next most recycled packaging material, in terms of weight recovered; 9.36 million tonnes of glass packaging waste were generated in the US in 2010, and 33.4% of the used material was recycled. Unlike paper-based material, the properties of glass are not affected by recycling. Cullet can be reprocessed into glass containers with physical properties identical to the original material. Glass recycling also reduces energy consumption: addition of 10% cullet reduces energy consumption by 2.5%. One of the major problems for food packaging applications originating from recycling is the presence of contaminants. Contaminants such as metals, papers, plastics, organic substances, ceramics, and heat-resistant glass (such as borosilicates), if not removed, can cause problems in making new food containers. Combination with colored glass is another problem, especially in the manufacture of colorless glass. This can result in off-color containers that are not acceptable for demanding applications such as food packaging. Mixed color cullet may be used for alternative applications such as abrasive paper, water filtration media, construction materials, and so on. Efficiently removing or sorting these contaminants or colored cullet requires a high level of investment. In addition, cullet is heavy and therefore expensive to ship. Collection within an acceptable transport distance is a critical point in glass recycling.

Recycling of Aluminum

Aluminum is a metal largely used for industrial and consumer goods applications; 1.9 million tonnes of total aluminum packaging waste were generated in the US in 2010, and 35.8% of the used material was recycled. Like glass, recycled aluminum does not lose its physical properties and is safe for food packaging. Using recycled aluminum can save 75–90% of energy compared with its production from natural sources. Collection of aluminum for recycling is through a combination of deposit, curbside, and dropoff programs. During the recycling process, aluminium is easily separated from other metals. Since it is lighter and not magnetic, iron or other ferrous metals can be separated using a magnetic separator or flotation. Internal coatings, printed ink, and any other organic contaminants are destroyed during the recycling process.

Recycling of Plastics

Plastics are recycled the least of the major packaging materials in the US; 13.68 million tonnes of plastic packaging waste was generated in the US in 2010, and 13.5% of the used material was recycled. During recycling, plastics can undergo several types of reactions such as chain scission, cross-linking, oxidation, and hydrolysis. Thus, the overall physical performance of the recycled plastics may decrease significantly. In addition, all recycled plastics cannot be mixed together due to their chemical incompatibility. In most applications, achieving good performance properties requires separation of the recycled plastics by resin type. Therefore, recycling of plastic packaging wastes is more difficult and costly than that of some other packaging materials. Energy recovery, such as incineration, may be the preferred option for multilayer plastics which cannot be separated by type, while cleaning and reprocessing into pellets for use in new plastic products is often the best choice for homogeneous plastics. Another option in some cases is depolymeriza- tion to monomers, purification, and repolymerization.

Since the potential migration of contaminants to food or other products is often a concern, most recycled plas- tics are used for non-food applications, including packag- ing containers, film and sheet, and also non-packaging applications such as fiber and carpet. In order to obtain food contact-grade plastic, the recycled plastics need to be processed and used in a way that effectively removes the potential for such contamination.

One option, as mentioned above, is chemical recycling, in which after depolymerization, the monomers are purified and repolymerized. Polymers produced in this way are identical to those produced from ordinary raw materials. Chemical recycling can be used for PET and PA, but is often not economical.

Another option for recycled plastic in food packaging is using it as a component in a multilayer structure. The recycled plastic is sandwiched inside, and a "barrier" layer of virgin plastic separates the recycled plastic from the product.

Another widely used option for PET is intensive clean- ing to remove most potential contaminants. A number of companies have received "non-objection" letters from the US FDA for plastics recycling processes that have been demonstrated to remove potential contaminants from recycled PET streams to a degree that makes them accept- able for food contact. One example is the Superclean process, which consists of a series of processes that can remove the volatile contaminants as well as increase the viscosity of the recycled PET so that it is suit- able for injection blow molding.

Most "non-objection" letters for recycled plastics have been issued for PET recycling processes. HDPE is both more susceptible to sorption of contaminants and more difficult to clean. While there have been a few processes approved for use of recycled HDPE in food packaging, there is little commercial use. On the other hand, recycled PET is used to a considerable extent in food packaging applications.

Biodegradable and Compostable Food Packaging

Due to growing concerns about waste disposal problems and the environmental effects of petroleum-based plastics, natural biopolymers derived from renewable sources that are biodegradable appear to be a good alternative to conventional plastics. In addition, in recent times, oil prices have increased markedly. These facts have caused increased interest in non-petroleum based biodegradable polymers. The biodegradation takes place through the action of enzymes and/or biochemical deterioration associated with living organisms, and the biodegradability depends not on the raw material sources to produce the polymer but rather on the chemical structure of the polymer and the environmental conditions because chemical structure, such as the chemical linkage, pending groups, etc., is related to susceptibility to degradation, and environmental condition is related to living organisms' activities.

The common challenges in using biodegradable pack- aging materials are ensuring durability to maintain their mechanical and barrier properties during the product's shelf life, and then, ideally, the ability to biodegrade quickly on disposal. Ideally, the materials need to function in a similar way to conventional packaging in filling and sealing equipment with equivalent costs. Bio based biodegradable polymers can be classified into three main categories according to their origin and method of production.

- Polymers directly extracted/removed from biomaterials (for example, starch, cellulose, casein, etc.).
- Polymers produced by classic chemical synthesis from monomers produced from biomaterials (for example, polylactide polymerized from lactic acid monomers).
- Polymers produced directly by microorganisms (for example, polyhydroxyalkanoates).

Biodegradable Polymers from Agricultural Crops

Polymers produced from starch are examples for this class. Starch is a widely available, environmentally friendly material with low cost. Corn is currently the most commonly used source of starch for bioplastics. However, potato, wheat, rice, barley, and oats can also be used as starch sources. Without modification, starch films are hydrophilic and have relatively poor mechanical strength. They cannot be used for packaging applications. The brittleness of starch-based bioplastics can be decreased by using biodegradable plasticizers, including glycerol and other low molecular weight polyhydroxy compounds, polyethers, etc.

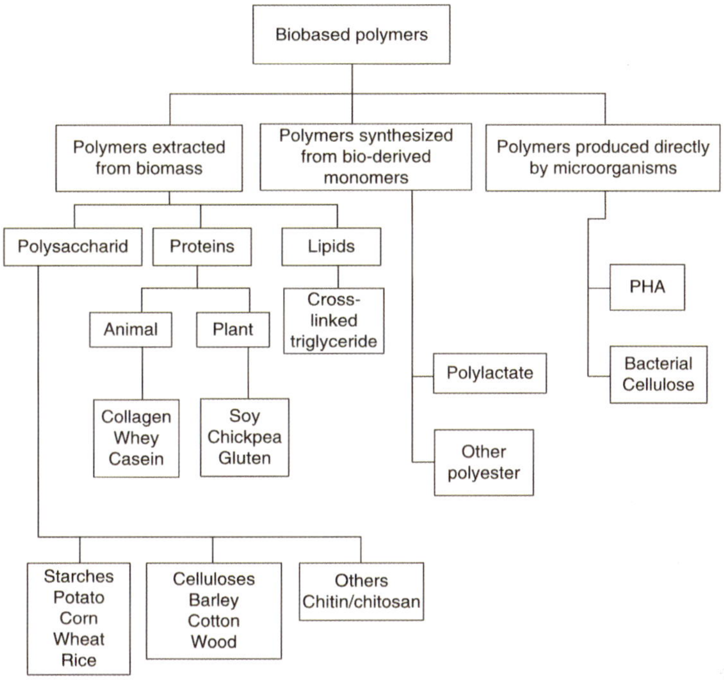

Figure: Biobased polymers for food packaging

Cellulose is the most widespread polysaccharide resource produced by plants. Cellulose is composed of glucose monomers that form a linear polymer with very long macromolecular chains. Cellulose is highly crystal- line, brittle, infusible, and insoluble in all organic solvents. These properties make cellulose impossible to process without modification. Cellophane films are produced by chemical modification of cellulose to render it soluble, and then regeneration of the cellulose after it is formed into film.

Another common practice is to use cellulose derivatives (cellulosic plastics) to improve the properties. The derivatives, such as ethers, esters and acetals, are produced by the reaction of one or more hydroxyl groups in the repeating unit, and can impart good film-forming properties. Cellulosic plastics are used for wrapping films, and for injection molded or blow molded containers

Biodegradable Polymers Synthesized from Bio-derived Monomers

Polylactic acid (PLA) is a biodegradable, thermoplastic polyester that is derived from lactic acid. Lactic acid can be produced economically by microbial fermentation of glucose obtained from biomaterials such as corn or wheat starch, lactose in whey, or sucrose. PLA is now produced on a comparatively large scale. PLA is usually obtained from ring-opening condensation of lactide, which is a lactic acid dimer. Two forms of the monomeric acid (D- or L-lactic acid) exist. The properties of PLAs can be varied by adjusting the relative amounts of the two lac- tic acid isomers (D- or L-) in the polymer. For example, 100% L-lactic acid or 100% D-lactic acid forms highly crystalline PLA whereas amorphous PLA is obtained from DL-co-polymers across a wide composition range. The L-isomer is predominant in most PLA resins as most lactic acid obtained from biological sources is the L form; 90%L/10%D PLA is a common formulation. The formation is able to crystallize but melts more easily than 100%L and is suitable for production of packaging films.

Different companies have commercialized PLA with various ratios of D/L lactide, with trade names including Nat- ureworks, Galacid, Heplon, etc.

Polylactic acids generally have reasonable moisture and oxygen barrier properties and are suitable for various plastic package forming processes such as blown and cast film, injection molding, and vacuum forming. PLA is currently utilized in wraps for bakery and confectionery products, paperboard coatings for cartons, disposable foodservice tableware items, containers for fresh produce, and water bottles. The rate of degradation of PLA depends on the degree of crystallinity. Increasing the amount of D-isomer in predominantly L-PLA tends to suppress crystallinity and therefore increase the rate of biodegradation. There has been some research on enhancing the biodegradability of PLA by grafting with chitosan.

Biodegradable Polymers Produced Directly by Microorganisms

Poly(β-hydroxyalcanoate)s (PHAs) are natural polyesters which are produced by bacteria from sugars or lipids. They are actually "grown" inside the cellular structure and then harvested. The use of PHAs is currently limited due to their high production costs. The performance properties are similar to conventional plastics, but the polymers are completely biodegradable. Thus, PHAs have potential as biodegradable alternative materials for conventional bulk commodity plastics. One of the family of PHAs, polyhydroxybutyrate (PHB), is the most commonly produced and researched bio- plastic. PHB has high thermal resistance and water barrier properties. However, a narrow processability window and low impact resistance have hampered widespread use of PHB for packaging application. Blends of PHB with other polymers may improve its properties. For example, poly(ethylene oxide), poly(vinyl butyral), poly(vinyl ace- tate), poly(vinylphenol), cellulose acetate butyrate, chitin, and chitosan have been studied as blend materials with PHB. Another common method to improve the processa- bility of PHB is to induce the microorganisms to form a co-polymer rather than the homopolymer. Polyhydroxy- butyrate-valerate, PHBV is the most common of these.

Synthetic Biodegradable Polymers

In addition to the biodegradable plastics based on natural substrates, there are synthetic biodegradable plastics pro- duced from petrochemical feedstocks that have groups. which are susceptible to hydrolytic microbial attack. Poly- caprolactone (PCL) is a semi-crystalline aliphatic polyes- ter which has a relatively low melting point (60° C). It is completely biodegradable in marine, sewage, sludge, soil, and compost ecosystems.

Polyvinyl alcohol (PVOH) is another synthetic biode-gradable polymer which is completely soluble in water. The combination of starch and PVOH as a biodegradable packaging material has been studied since 1970. Currently, it is used to produce starch-based loose fillers as a substitute for expanded PS. Other types of synthetic biodegradable polymers include polyesters, polyamides, polyurethanes and polyureas, poly(amide-enamine)s, polyanhydrides.

Synthetic polymers can be manipulated for a wide range of properties to obtain required mechanical properties (flexibility, toughness, etc.) as well as the degree of degradation. Rather than food packaging, the application of synthetic biodegradable polymers has been gaining more attention in the biomedical area such as tissue engineering scaffolds, orthopedic fixation devices, etc.

Sanitizers and Disinfectants

In the food industry, chemicals are routinely used to sanitize and disinfect product contact surfaces. These chemicals provide a necessary and required step to ensure that the foods produced and consumed are as free as possible from microorganisms that can cause foodborne illness. Prevention is the name of the game. What are these chemicals, how do they function and how are they used?

Disinfecting Versus Sanitizing

Before discussing the chemicals, the differences between sanitizers and disinfectants as used in the food industry must be understood. To disinfect means to destroy or irreversibly inactivate specified infectious fungi and bacteria, but not necessarily the spores, on hard surfaces. To sanitize means to reduce microorganisms of public health importance to levels considered safe, based on established parameters, without adversely affecting either the quality of the product or its safety. While disinfection measures may be employed in food processing and preparation, it is much more common to utilize sanitization methods to reduce microbial presence.

To achieve the required level of sanitization or disinfection, the chemical in question must be applied at a certain concentration for a specified amount of time. These parameters are described on the product label and must be followed to achieve the desired microbial control. In most cases, these products are registered for use as pesticides with the U.S. Environmental Protection Agency (EPA). Once applied, the allowable residues and the monitoring thereof in food processing and preparation are the responsibility of the U.S. Food and Drug Administration. Facilities operating under the jurisdiction of the U.S. Department of Agriculture must additionally use products approved by that agency. Of course, the task of ensuring the chemicals are prepared and applied properly to avoid inappropriate residues rests with the food processor and foodservice operator.

The efficacy of a chemical used for sanitizing or disinfection rests upon its ability to reduce the contamination level. The sanitization standard for contamination reduction of food contact surfaces is generally accepted as 99.999% (a 5-log reduction) achieved in 30 seconds (Official Detergent Sanitizer Test). The sanitization standard for non-food contact surfaces is accepted as a reduction of 99.9% (3 logs) within 30 seconds. Disinfection, in contrast, must destroy or irreversibly inactivate all specified organisms within a certain time, usually 10 minutes. Some chemicals may function as both sanitizers and disinfectants.

The process of sanitization depends upon the preparation of the surfaces in question. Most sanitizers must be applied to surfaces that are free of organic matter and cleaner residues. The generally accepted order of events is rinse, clean, rinse and sanitize. The cleaner utilized in the cleaning step needs to be oriented and appropriate for the soil present. For example, alkaline detergents more efficiently remove fat- and protein-based soils, while mineral-based soils require acid cleaners. Thankfully, modern cleaning agents are mixtures of chemical components that can address various cleaning scenarios.

Sanitizing Chemicals

The food industry most often uses sanitizing procedures, so the information presented herein will focus on the more common products utilized. Regardless of the product, the sanitizing solution must be tested to verify that the desired concentration is consistently present. Too little sanitizer, of course, can result in unacceptable efficacy, while too much sanitizer can yield residues that do not meet standards.

Hypochlorites

Effectiveness, low cost and ease of manufacturing make hypochlorites the most widely used sanitizers. Sodium hypochlorite is the most common compound and is an ideal sanitizer, as it is a strong oxidizer.

Hypochlorites cause broad microbial mortality by damaging the outer membrane, likely producing a loss of permeability control and eventual lysis of the cell. In addition, these compounds inhibit cellular enzymes and destroy DNA. Spores, however, are resistant to hypochlorites, as the spore coat is not susceptible to oxidation except at high concentrations coupled with long contact times at elevated temperatures.

While hypochlorites are very reactive, their useful properties are negatively impacted by factors such as suspended solids, high temperatures, light, water impurities and improper pH levels. In routine use, surfaces must be as free as possible of organic materials, and the pH must be maintained between 5 to 7 to ensure that the greatest amount of hypochlorous acid is available. As with any sanitizer, measurements must be taken periodically to make certain that the freely available chlorine is at the desired level. For no-rinse applications, the maximum allowable concentration of available chlorine is 200 ppm.

Other disadvantages of hypochlorites are corrosiveness to metals, health concerns related to skin irritation and mucous membrane damage and environmental contamination. The latter is of concern as chlorine can combine with organic substances to form toxic chlorinated compounds, such as trihalomethanes and dioxins. Hypochlorite use may be further restricted in the future. Care must be taken when cleaning hypochlorite spills as organic materials such as cloth, sawdust and paper may spontaneously combust upon drying.

Chlorine Dioxide

This inorganic compound is a broad sanitizer effective against bacteria, fungi and viruses. Chlorine dioxide is an oxidizer that reacts with the proteins and fatty acids within the cell membrane, resulting in loss of permeability control and disruption of protein synthesis.

While chlorine dioxide is an explosive gas, it is relatively safe in solution. It is produced on-site as it can't be compressed or stored commercially in gaseous form. Most chlorine dioxide generation is accomplished with complex systems. However, recent advances in formulation procedures allow the production of solutions of chlorine dioxide on-site without the use of expensive equipment.

Compared with hypochlorites, chlorine dioxide requires much lower concentrations to achieve microbial mortality. For example, a 5-ppm solution is effective as a sanitizer on food contact surfaces with a contact time of at least 1 minute. Further, disinfection can be achieved with 100 ppm using a contact time of 10 minutes.

Chlorine dioxide reacts more selectively with compounds present in microbial cells as opposed to reacting with organic compounds in general. This ability allows chlorine dioxide to function in more organically loaded solutions, though as organic load increases, efficacy does decrease. Chlorine dioxide functions well over a pH range of about 6 to 10, thus allowing increased mortality of some microbes at higher values. Another advantage is that chlorine dioxide does not form chlorinated organic compounds, making it more environmentally friendly.

Iodophors

These compounds are less active than hypochlorites but are effective sanitizers and disinfectants. Iodophors attach to the sulfurs of proteins such as cysteine, causing inactivation and cell wall damage. Carriers with iodophor solutions allow a sustained-release effect, resulting in continuous microbial mortality.

Iodophors fare better in situations in which the pH is slightly acidic, as less active forms exist above neutral pH. The common concentration for sanitization is 25 ppm for 1 minute. Unfortunately, iodine compounds easily stain many surfaces, particularly plastics. On the plus side, they are common sanitizers used on glass surfaces, such as in the beer and wine bottling industries. The EPA has assessed iodophors as having no significant effect on the environment.

Peroxyacetic Acid (PAA)

PAA is an effective sanitizer that is active against many microorganisms and their spores. Mortality is produced by the disruption of chemical bonds within the cell membrane. PAA-based sanitizers are frequently paired with stabilized hydrogen peroxide. These sanitizers function well under cold conditions (~ 4° C), thus producing acceptable microbial mortality on equipment normally held below ambient temperature. PAA is also effective in removing biofilms and is more active than hypochlorites.

PAA solutions can be attenuated by the organic load and will begin to lose activity as the pH approaches neutral. These solutions are applied at concentrations ranging from about 100 ppm to 200 ppm for peroxyacetic acid and 80 ppm to 600 ppm for hydrogen peroxide.

PAA-based sanitizers are environmentally friendly as the compounds therein break down into acetic acid, oxygen and water. These sanitizers are also less corrosive to equipment than hypochlorites. As with any highly active oxidizer, concentrated PAA can present a safety hazard.

Quaternary Ammonium Compounds (QACs or Quats)

Quaternary ammonium compounds are fairly complex chemicals in which nitrogen is bound to four organic groups. The positively charged cations in the compounds bind with the acidic phospholipids in the microbial cell wall. This action blocks the uptake of nutrients into the microbial cell and prevents the discharge of waste. In general, QACs are effective against a wide range of microbes, although the spore phase is unaffected. At lower concentrations, Gram-positive bacteria are more sensitive to QACs than Gram-negative bacteria. QACs are formulated in many different variations for specific situations.

QACs may be applied at concentrations varying from about 100 ppm to 400 ppm. As sanitizers, QACs are commonly applied at 200 ppm to food contact surfaces, and the solution is allowed to dry. Once dry, a residue of the QAC compounds remains and provides germicidal activity until degradation occurs. QACs also can function as detergents when present in high concentration because the compounds possess both hydrophilic and lipophilic chemical groups.

QACs are usually odorless, nonstaining, noncorrosive and relatively nontoxic to users. They function well over a broad temperature range and a wide pH range, although activity is greater at warmer temperatures and in alkaline situations. While QACs tolerate light organic loads, heavy soil will decrease QAC activity significantly. Some QACs may not function adequately in hard water, but others are formulated with added chelating agents that allow such use.

While QACs do combine with organic compounds and are discharged into the environment, the concentrations are low and heterotrophic bacteria are not negatively impacted. Soil-inhabiting bacteria such as *Pseudomonas* spp. and *Xanthomonas* spp. can degrade QACs. In addition, the low amounts of QACs flowing into commercial sewage treatment facilities appear to combine with the anionic surfactants present to form complexes that reduce or eliminate toxicity.

Resistance to Sanitizers

Any time a chemical is used to produce microbial mortality, the possibility of promoting resistance exists. This is because not all of the microbes are killed. A 5-log reduction (99.999%) still means that of 1,000,000 microbes present, 10 have survived, even though the process has reduced the population to what can be termed a safe level. The sanitizer could have just missed these 10 organisms or they could inherently be immune. If these 10 microbes are indeed immune, over time they will proliferate, and the usual sanitizing concentration and chemical will no longer produce acceptable mortality. At this point, measures must be taken to disinfect the surfaces in question. It then becomes imperative to know what organisms are specifically present so that the proper disinfectant at the proper strength maintained for the required time can be applied.

Sometimes, it is thought that bacterial resistance is present when actually the organisms are avoiding contact with the sanitizing chemical because a biofilm is present. Biofilms are polysaccharides that allow attachment to most any surface. Bacteria such as *Escherichia coli, Salmonella* spp., *Listeria* spp., *Campylobacter* spp. and several others can produce biofilms. Over time, the film becomes enhanced and may contain different species of bacteria, yielding a constant source of contamination. Whether biofilms are truly products of bacterial resistance may be a philosophical question, but the presence of biofilms can be an indisputable problem for those assigned to find a solution.

Removal of Microorganisms

Removal of microorganisms is not very effective in food preservation but under special conditions it may be helpful. Removal may be accomplished by means of filtration, centrifugation, washing, trimming.

- Filtration:

It is the only successful method for the complete removal of organisms and its use is limited to clear liquids. This method has been used successfully with fruit juices, beer, soft drinks, wine and water.

- Centrifugation:

It is not very effective. All microorganisms are not removed. Ex: Used in the treatment of drinking water but it is insufficient to remove micro organisms.

- In the milk, main purpose of centrifugation is not to remove bacteria but to take out other suspended materials, but bactofugation is to remove bacteria from milk.

- Washing:

Washing raw foods remove spoilage microorganisms. Ex: Cabbage heads or cucumbers before their fermentation into sauerkraut and pickles, removes most of the soil microorganisms on the surface.

Washing fresh fruits and vegetables removes soil organisms that may be resistant to the heat process during canning.

Washing foods is dangerous if water adds spoilage microorganisms and increases the moisture so that growth of spoilage organisms is encouraged.

- Trimming:

Spoiled portions of a food removed by trimming. Large no. of spoilage organisms are removed by this way.

Ex: Trimming the outer leaves of cabbage heads for the manufacture of sauerkraut.

Artificial Waxing

Coating fruits and vegetables with wax have become a common practice to improve post-harvest consumer appeal. The fruit waxing technique, also known as artificial waxing, gives fruits their artificial freshness and cleanliness. Some wax coating helps to retain the moisture in the fruits and also inhibits mould growth. Recently, there have been a lot of controversies on using fruit waxing and eating these wax-coated fruits that leaves the consumers in a total dilemma.

Wax-coated fruits may appear fresh, have a long shelf life, but can also have harmful effects on the human health.

Fruit waxing is a technique of coating fruits with either natural edible wax or artificial wax. Fruit waxing leaves a thin shiny layer on the surface of fruit in addition to or as a replacement of natural wax produced by the fruit itself. The main objective of fruit waxing is to provide a barrier to moisture loss, prevent bruising, prevent physical damage, and delay browning of fruits.

Apple is one of the most common fruits that are covered with artificial wax. Wax coated fruits are appealing to consumers as these appear fresh and clean compared to uncoated fruits that show browning and lose their firmness quickly. Fruit waxing reduces the loss of water from fruits and thus preserves texture and flavor. Wax-coated fruits also show slow ripening as the wax acts as a barrier to environmental gases and water vapour.

Need of Artificial Waxing Coated on Fruits

Many fruits and vegetables produce their own natural protective wax coating called cuticle that consists of substances such as fats, resins, or waxes. This natural wax coating on fruits helps to prevent moisture loss from inside fruit flesh. However, the natural protective layer easily wears off after the fruit is plucked from the tree and during the process of packing and transportation.

Therefore, to retain the moisture in the fruit and make it appear as fresh as farm picked till it reaches the consumer, most fruits are covered with artificial waxes. The fruit wax coatings may be composed of polysaccharides, proteins, lipids, and composites. The fruit waxing itself does not control decay of fruits and hence the wax coating is combined with some chemicals.

Fruit Waxing Methods

The fruit waxing method can be manual or automated. During manufacturing, the waxy coating material is dispersed and dissolved in a solvent such as water, alcohol, a mixture of water and alcohol.

Additives such as plasticizers, antimicrobial agents, minerals, vitamins, colors, or flavors can be added in this process. The film solutions can be applied to fruits by several methods such as dipping, spraying, brushing, and panning followed by drying.

- Manual Rubbing

  The process is performed by applying the wax coating over fruits surface using a brush with soft bristles or absorbent cloth. After application of edible wax, the fruits are air dried for about 15 minutes. It is a time-consuming process and also requires manpower.

- Dipping

  The dipping method for fruit waxing is used when the fruits are coated with paraffin wax.

  The fruits are dipped in melted paraffin wax coat for about a second. The paraffin solidifies immediately on the fruit's surface after it is removed from the wax bath.

- Brushing

  This is an automated method of fruit waxing. The liquefied wax is dispensed over the brush that continuously applies a thin layer of wax coating over fruits surface. For even application of edible wax, the brush needs to be completely saturated with the wax. The wax can be sprayed using a pump through low-pressure nozzles.

Safety of Wax-coated Fruits

Whether wax-coated fruits are safe to eat or not totally depends on the kind of wax coating used. Research by some scientific authorities suggests that food-grade wax is safe to eat. These are the edible waxes that are considered safe as they are not broken down by the body for absorption and are eliminated by the body.

However, most fruit waxes are mixed with morpholine and its derivatives (MAID) to ensure that they are applied thinly and evenly. In the presence of nitrate, contained in diet, morpholine can be chemically nitrosated to form N-nitrosomorpholine (NMOR) a potent carcinogenic compound.

Morpholine is also believed to cause impaired liver or kidney function. Morpholine at the levels 0.03-0.3ppm, is likely to be of low-risk to human health. However, above this level morpholine can be toxic and unsafe. Chronic consumption of wax-coated fruits and vegetables containing morpholine could present a slight risk for cancer in certain individuals.

Harmful Effects

The fruit waxes are mixed with additives such as glycerols, lactic acid or acetic acid to adjust the pH of coating material. In addition, the fruit waxes may also contain traces of preservatives, antimicrobial agents, and texture enhancers.

Morpholine is present in most fruit waxes to give thin and even films. The safe dose of morpholine in humans is 4.3ng/kg body weight/day. Repetitive consumption of morpholine as well as some other chemical agents present in fruit waxing can be hazardous to health.

The health risks associated with regular consumption of artificially wax coated fruits include,

- Risk of Cancer

  Morpholine is used commonly as a solvent and emulsifier in making the wax coatings for fruits and vegetables. Morpholine, by itself, in the doses that are present in fruits and vegetables does not constitute a health risk.

  However, inside the body when it comes in contact with nitrate, it forms Nitrosomorpholine (NMOR), a genotoxic carcinogen that poses a risk of liver or kidney cancer.

- Risk of Liver and Kidney Damage

  Experts report that on oral and parenteral administration or after inhalation, morpholine is well absorbed and is distributed in the body fluids. Ingestion of morpholine through daily consumption of wax-coated fruits can affect liver and kidney function.

- Allergies

  Many edible coatings are made from ingredients that could cause allergic reactions. These allergens include protein substances such as soy, whey protein, casein, and peanut proteins.

Tips to Prevent Health Hazards of Wax-coated Fruits

The use of morpholine as a glazing agent in fruit waxing is common. Although it is present in negligible amounts in the waxy coating, accumulation of NMOR by daily intake of coated fruits/ vegetables is a serious threat to health. To overcome this problem and prevent the health hazards of fruit waxing follow these tips given below:

- Follow the mantra of 3ws: Wipe, wash and soak the wax-coated fruits in warm water for a longer time. You may also scrub the skin with a brush to get rid of edible wax coating on fruits.
- Don't fall prey to artificial freshness: Do not get influenced by bright colors of fruits and vegetables that appear shiny.
- Buy more local and seasonal fruits: Fruits that are available off season may be the ones that are stored for a long time with the help of wax coatings. Try to buy fruits from local farmer or buy fruits at the local market that are in the season.
- Read the labels carefully: The FDA makes it mandatory for the packers and providers to mention the presence of wax coat as 'food grade animal or vegetable wax'. You may also look for fruits labeled as 'No wax or resin coating'.
- Prefer natural edible wax: Look for fruits that have a natural wax coating such as aloe vera gel or Chitosan, which do not contain amines and are also natural alternatives to synthetic coating.
- Eat fruits without skin: If you are too concerned about health hazards of fruit waxing peel the skin of the fruit before you eat.

Preservation

Food preservation includes all the techniques which are used to prevent food from spoiling. It includes methods such as canning, pickling, drying and freeze-drying, irradiation, pasteurization, smoking, and the addition of chemical additives. Food preservation has become an increasingly important component of the food industry as fewer people eat foods produced on their own lands, and as consumers expect to be able to purchase and consume foods that are out of season.

The vast majority of instances of food spoilage can be attributed to one of two major causes:

(1) The attack by pathogens (disease-causing microorganisms) such as bacteria and molds,

(2) Oxidation that causes the destruction of essential biochemical compounds and the destruction of plant and animal cells.

The various methods that have been devised for preserving foods are all designed to reduce or eliminate one or the other (or both) of these causative agents.

For example, a simple and common method of preserving food is by heating it to some minimum temperature. This process prevents or retards spoilage because high temperatures kill or inactivate most kinds of pathogens. The addition of compounds known as BHA and BHT to foods also prevents spoilage in another different way. These compounds are known to act as antioxidants, preventing chemical reactions that cause the oxidation of food that results in its spoilage. Almost all techniques of preservation are designed to extend the life of food by acting in one of these two ways.

The search for methods of food preservation probably can be traced to the dawn of human civilization. People who lived through harsh winters found it necessary to find some means of insuring a food supply during seasons when no fresh fruits and vegetables were available. Evidence for the use of dehydration (drying) as a method of food preservation, for example, goes back at least 5,000 years. Among the most primitive forms of food preservation that are still in use today are such methods as smoking, drying, salting, freezing, and fermenting.

Early humans probably discovered by accident that certain foods exposed to smoke seem to last longer than those that are not. Meats, fish, fowl, and cheese were among such foods. It appears that compounds present in wood smoke have antimicrobial actions that prevent the growth of organisms that cause spoilage. Today, the process of smoking has become a sophisticated method of food preservation with both hot and cold forms in use. Hot smoking is used primarily with fresh or frozen foods, while cold smoking is used most often with salted products. The most advantageous conditions for each kind of smoking—air velocity, relative humidity, length of exposure, and salt content, for example—are now generally understood and applied during the smoking process. For example, electrostatic precipitators can be employed to attract smoke particles and improve the penetration of the particles into meat or fish. So many alternative forms of preservation are now available that smoking no longer holds the position of importance it once did with ancient peoples. More frequently, the process is used to add interesting and distinctive flavors to foods.

Because most disease-causing organisms require a moist environment in which to survive and multiply, drying is a natural technique for preventing spoilage. Indeed, the act of simply leaving foods out in the sun and wind to dry out is probably one of the earliest forms of food preservation. Evidence for the drying of meats, fish, fruits, and vegetables go back to the earliest recorded human history. At some point, humans also learned that the drying process could be hastened and improved by various mechanical techniques. For example, the Arabs learned early on that apricots could be preserved almost indefinitely by macerating them, boiling them, and then leaving them to dry on broad sheets. The product of this technique, quamaradeen, is still made by the same process in modern Muslim countries.

Today, a host of dehydrating techniques are known and used. The specific technique adopted depends on the properties of the food being preserved. For example, a traditional method for preserving rice is to allow it to dry naturally in the fields or on drying racks in barns for about two weeks. After this period of time, the native rice is threshed and then dried again by allowing it to sit on straw mats in the sun for about three days. Modern drying techniques make use of fans and heaters in controlled environments. Such methods avoid the uncertainties that arise from leaving crops in the field to dry under natural conditions. Controlled temperature air drying is especially popular for the preservation of grains such as maize, barley, and bulgur.

Vacuum drying is a form of preservation in which a food is placed in a large container from which air is removed. Water vapor pressure within the food is greater than that outside of it, and water evaporates more quickly from the food than in a normal atmosphere. Vacuum drying is biologically desirable since some enzymes that cause oxidation of foods become active during normal air drying. These enzymes do not appear to be as active under vacuum drying conditions, however. Two of the special advantages of vacuum drying are that the process is more efficient at removing water from a food product, and it takes place more quickly than air drying. In one study, for example, the drying time of a fish fillet was reduced from about 16 hours by air drying to six hours as a result of vacuum drying.

Coffee drinkers are familiar with the process of dehydration known as spray drying. In this process, a concentrated solution of coffee in water is sprayed though a disk with many small holes in it. The surface area of the original coffee grounds is increased many times, making dehydration of the dry product much more efficient. Freeze-drying is a method of preservation that makes use of the physical principle known as sublimation. Sublimation is the process by which a solid passes directly to the gaseous phase without first melting. Freeze-drying is a desirable way of preserving food because at low temperatures (commonly around 14° F to –13° F [–10° C to –25° C]) chemical reactions take place very slowly and pathogens have difficulty surviving. The food to be preserved by this method is first frozen and then placed into a vacuum chamber. Water in the food first freezes and then sublimes, leaving a moisture content in the final product of as low as 0.5%.

The precise mechanism by which salting preserves food is not entirely understood. It is known that salt binds with water molecules and thus acts as a dehydrating agent in foods. A high level of salinity may also impair the conditions under which pathogens can survive. In any case, the value of adding salt to foods for preservation has been well known for centuries. Sugar appears to have effects similar to those of salt in preventing spoilage of food. The use of either compound (and of certain other natural materials) is known as curing. A desirable side effect of using salt or sugar as a food preservative is, of course, the pleasant flavor each compound adds to the final product.

Curing can be accomplished in a variety of ways. Meats can be submerged in a salt solution known as brine, for example, or the salt can be rubbed on the meat by hand. The injection of salt solutions into meats has also become popular. Food scientists have now learned that a number of factors relating to the food product and to the preservative conditions affect the efficiency of curing. Some of the food factors include the type of food being preserved, the fat content, and the size of treated pieces. Preservative factors include brine temperature and concentration, and the presence of impurities.

Curing is used with certain fruits and vegetables, such as cabbage (in the making of sauerkraut), cucumbers (in the making of pickles), and olives. It is probably most popular, however, in the preservation of meats and fish. Honey-cured hams, bacon, and corned beef ("corn" is a term for a form of salt crystals) are common examples.

Freezing is an effective form of food preservation because the pathogens that cause food spoilage are killed or do not grow very rapidly at reduced temperatures. The process is less effective in food preservation than are thermal techniques such as boiling because pathogens are more likely to be able to survive cold temperatures than hot temperatures. In fact, one of the problems surrounding the use of freezing as a method of food preservation is the danger that pathogens deactivated (but not killed) by the process will once again become active when the frozen food thaws.

A number of factors are involved in the selection of the best approach to the freezing of foods, including the temperature to be used, the rate at which freezing is to take place, and the actual method used to freeze the food. Because of differences in cellular composition, foods actually begin to freeze at different temperatures ranging from about 31° F (−0.6° C) for some kinds of fish to 19° F (−7° C) for some kinds of fruits.

The rate at which food is frozen is also a factor, primarily because of aesthetic reasons. The more slowly food is frozen, the larger the ice crystals that are formed. Large ice crystals have the tendency to cause rupture of cells and the destruction of texture in meats, fish, vegetables, and fruits. In order to deal with this problem, the technique of quick-freezing has been developed. In quick-freezing, a food is cooled to or below its freezing point as quickly as possible. The product thus obtained, when thawed, tends to have a firm, more natural texture than is the case with most slow-frozen foods.

About a half dozen methods for the freezing of foods have been developed. One, described as the plate, or contact, freezing technique, was invented by the American inventor Charles Birdseye in 1929. In this method, food to be frozen is placed on a refrigerated plate and cooled to a temperature less than its freezing point. Alternatively, the food may be placed between two parallel refrigerated plates and frozen. Another technique for freezing foods is by immersion in very cold liquids. At one time, sodium chloride brine solutions were widely used for this purpose. A 10% brine solution, for example, has a freezing point of about 21° F (−6° C), well within the desired freezing range for many foods. More recently, liquid nitrogen has been used for immersion freezing. The temperature of liquid nitrogen is about −320° F (−195.5° C), so that foods immersed in this substance freeze very quickly.

As with most methods of food preservation, freezing works better with some foods than with others. Fish, meat, poultry, and citrus fruit juices (such as frozen orange juice concentrate) are among the foods most commonly preserved by this method.

Fermentation is a naturally occurring chemical reaction by which a natural food is converted into another form by pathogens. It is a process in which food spoils, but results in the formation of an edible product. Perhaps the best example of such a food is cheese. Fresh milk does not remain in edible condition for a very long period of time. Its pH is such that harmful pathogens begin to grow in it very rapidly. Early humans discovered, however, that the spoilage of milk can be controlled in such a way as to produce a new product, cheese.

Bread is another food product made by the process of fermentation. Flour, water, sugar, milk, and other raw materials are mixed together with yeasts and then baked. The addition of yeasts brings about the fermentation of sugars present in the mixture, resulting in the formation of a product that will remain edible much longer than will the original raw materials used in the bread-making process.

Heating food is an effective way of preserving it because the great majority of harmful pathogens are killed at temperatures close to the boiling point of water. In this respect, heating foods is a form of food preservation comparable to that of freezing but much superior to it in its effectiveness. A preliminary step in many other forms of food preservation, especially forms that make use of packaging, is to heat the foods to temperatures sufficiently high to destroy pathogens.

In many cases, foods are actually cooked prior to their being packaged and stored. In other cases, cooking is neither appropriate nor necessary. The most familiar example of the latter situation is pasteurization. During the 1860s, the French bacteriologist Louis Pasteur discovered that pathogens in foods could be destroyed by heating those foods to a certain minimum temperature. The process was particularly appealing for the preservation of milk since preserving milk by boiling is not a practical approach. Conventional methods of pasteurization called for the heating of milk to a temperature between 145 and 149° F (63 and 65° C) for a period of about 30 minutes, and then cooling it to room temperature. In a more recent revision of that process, milk can also be "flash-pasteurized" by raising its temperature to about 160°F (71° C) for a minimum of 15 seconds, with equally successful results. A process known as ultra-high-pasteurization uses even higher temperatures, of the order of 194–266° F (90–130° C), for periods of a second or more.

One of the most common methods for preserving foods today is to enclose them in a sterile container. The term "canning" refers to this method although the specific container can be glass, plastic, or some other material as well as a metal can, from which the procedure originally obtained its name. The basic principle behind canning is that a food is sterilized, usually by heating, and then placed within an air-tight container. In the absence of air, no new pathogens can gain access to the sterilized food. In most canning operations, the food to be packaged is first prepared in some way—cleaned, peeled, sliced, chopped, or treated in some other way—and then placed directly into the container. The container is then placed in hot water or some other environment where its temperature is raised above the boiling point of water for some period of time. This heating process achieves two goals at once. First, it kills the vast majority of pathogens that may be present in the container. Second, it forces out most of the air above the food in the container.

After heating has been completed, the top of the container is sealed. In home canning procedures, one way of sealing the (usually glass) container is to place a layer of melted paraffin directly on top of the food. As the paraffin cools, it forms a tight solid seal on top of the food. Instead of or in addition to the paraffin seal, the container is also sealed with a metal screw top containing a rubber gasket.

The commercial packaging of foods frequently makes use of tin, aluminum, or other kinds of metallic cans. The technology for this kind of canning was first developed in the mid-1800s, when individual workers hand-sealed cans after foods had been cooked within them. At this stage, a single worker could seldom produce more than 100 "canisters" (from which the word "can" later came) of food a day. With the development of far more efficient canning machines in the late nineteenth century, the mass production of canned foods became a reality.

As with home canning, the process of preserving foods in metal cans is simple in concept. The foods are prepared and the empty cans are sterilized. The prepared foods are then added to the sterile metal can, the filled can is heated to a sterilizing temperature, and the cans are then sealed by a machine. Modern machines are capable of moving a minimum of 1,000 cans per minute through the sealing operation.

The majority of food preservation operations used today also employ some kind of chemical additive to reduce spoilage. Of the many dozens of chemical additives available, all are designed either to kill or retard the growth of pathogens or to prevent or retard chemical reactions that result in the oxidation of foods. Some familiar examples of the former class of food additives are sodium benzoate and benzoic acid; calcium, sodium propionate, and propionic acid; calcium, potassium, sodium sorbate, and sorbic acid; and sodium and potassium sulfite. Examples of the latter class of additives include calcium, sodium ascorbate, and ascorbic acid (vitamin C); butylated hydroxyanisole (BHA) and butylated hydroxytoluene (BHT); lecithin; and sodium and potassium sulfite and sulfur dioxide.

A special class of additives that reduce oxidation is known as the sequestrants. Sequestrants are compounds that "capture" metallic ions, such as those of copper, iron, and nickel, and remove them from contact with foods. The removal of these ions helps preserve foods because in their free state they increase the rate at which oxidation of foods takes place. Some examples of sequestrants used as food preservatives are ethylenediamine-tetraacetic acid (EDTA), citric acid, sorbitol, and tartaric acid.

The lethal effects of radiation on pathogens has been known for many years. Since the 1950s, research in the United States has been directed at the use of this technique for preserving certain kinds of food. The radiation used for food preservation is normally gamma radiation from radioactive isotopes or machine-generated x rays or electron beams. One of the first applications of radiation for food preservation was in the treatment of various kinds of herbs and spices, an application approved by the U.S. Food and Drug Administration (FDA) in 1983. In 1985, the FDA extended its approval to the use of radiation for the treatment of pork as a means of destroying the pathogens that cause trichinosis. Experts predict that the ease and efficiency of food preservation by means of radiation will develop considerably in the future. That future is somewhat clouded, however, by fears expressed by some scientists and members of the general public about the dangers that irradiated foods may have for humans. In addition to a generalized concern about the possibilities of being exposed to additional levels of radiation in irradiated foods (not a possibility), critics have raised questions about the creation of new and possibly harmful compounds in food that has been exposed to radiation.

Importance of Food Preservation

Food preservation helps in:

- Increasing the self-life of foods thus increasing the supply. So many perishable foods can be preserved for a long time.
- Making the seasonal food available throughout the year.
- Adding variety to the diet.
- Saving time by reducing preparation time and energy, as the food has already been partially processed.
- Stabilising prices of food, as there is less scope of shortage of supply to demand.
- Decreasing wastage of food by preventing decay or spoilage of food.

- Improving the nutrition of the population. Preserved foods help people to bring a variety in the diet, thereby decreasing nutritional inadequacies.

Food Additives

Food additive is any of various chemical substances added to foods to produce specific desirable effects. Additives such as salt, spices, and sulfites have been used since ancient times to preserve foods and make them more palatable. With the increased processing of foods in the 20th century, there came a need for both the greater use of and new types of food additives. Many modern products, such as low-calorie, snack, and ready-to-eat convenience foods, would not be possible without food additives.

There are four general categories of food additives: nutritional additives, processing agents, preservatives, and sensory agents. These are not strict classifications, as many additives fall into more than one category.

Nutritional Additives

Nutritional additives are used for the purpose of restoring nutrients lost or degraded during production, fortifying or enriching certain foods in order to correct dietary deficiencies, or adding nutrients to food substitutes. The fortification of foods began in 1924 when iodine was added to table salt for the prevention of goitre. Vitamins are commonly added to many foods in order to enrich their nutritional value. For example, vitamins A and D are added to dairy and cereal products, several of the B vitamins are added to flour, cereals, baked goods, and pasta, and vitamin C is added to fruit beverages, cereals, dairy products, and confectioneries. Other nutritional additives include the essential fatty acid linoleic acid, minerals such as calcium and iron, and dietary fibre.

Processing Agents

A number of agents are added to foods in order to aid in processing or to maintain the desired consistency of the product.

Processing additives and their uses		
function	typical chemical agent	typical product
anticaking	sodium aluminosilicate	salt
bleaching	benzoyl peroxide	flour
chelating	ethylenediaminetetraacetic acid (EDTA)	dressings, mayonnaise, sauces, dried bananas
clarifying	bentonite, proteins	fruit juices, wines
conditioning	potassium bromate	flour
emulsifying	lecithin	ice cream, mayonnaise, bakery products
leavening	yeast, baking powder, baking soda	bakery products
moisture control (humectants)	glycerol	marshmallows, soft candies, chewing gum

pH control	citric acid, lactic acid	certain cheeses, confections, jams and jellies
stabilizing and thickening	pectin, gelatin, carrageenan, gums (arabic, guar, locust bean)	dressings, frozen desserts, confections, pudding mixes, jams and jellies

Emulsifiers are used to maintain a uniform dispersion of one liquid in another, such as oil in water. The basic structure of an emulsifying agent includes a hydrophobic portion, usually a long-chain fatty acid, and a hydrophilic portion that may be either charged or uncharged. The hydrophobic portion of the emulsifier dissolves in the oil phase, and the hydrophilic portion dissolves in the aqueous phase, forming a dispersion of small oil droplets. Emulsifiers thus form and stabilize oil-in-water emulsions (e.g., mayonnaise), uniformly disperse oil-soluble flavour compounds throughout a product, prevent large ice crystal formation in frozen products (e.g., ice cream), and improve the volume, uniformity, and fineness of baked products.

Stabilizers and thickeners have many functions in foods. Most stabilizing and thickening agents are polysaccharides, such as starches or gums, or proteins, such as gelatin. The primary function of these compounds is to act as thickening or gelling agents that increase the viscosity of the final product. These agents stabilize emulsions, either by adsorbing to the outer surface of oil droplets or by increasing the viscosity of the water phase. Thus, they prevent the coalescence of the oil droplets, promoting the separation of the oil phase from the aqueous phase (i.e., creaming). The formation and stabilization of foam in a food product occurs by a similar mechanism, except that the oil phase is replaced by a gas phase. The compounds also act to inhibit the formation of ice or sugar crystals in foods and can be used to encapsulate flavour compounds.

Chelating, or sequestering, agents protect food products from many enzymatic reactions that promote deterioration during processing and storage. These agents bind to many of the minerals that are present in food (e.g., calcium and magnesium) and are required as cofactors for the activity of certain enzymes.

Preservatives

Food preservatives are classified into two main groups: antioxidants and antimicrobials. Antioxidants are compounds that delay or prevent the deterioration of foods by oxidative mechanisms. Antimicrobial agents inhibit the growth of spoilage and pathogenic microorganisms in food.

Food preservatives	
chemical agent	mechanism of action
Antioxidants	
ascorbic acid	oxygen scavenger
butylated hydroxyanisole (BHA)	free radical scavenger
butylated hydroxytoluene (BHT)	free radical scavenger
citric acid	enzyme inhibitor/metal chelator

sulfites	enzyme inhibitor/oxygen scavenger
tertiary butylhydroquinone (TBHQ)	free radical scavenger
tocopherols	free radical scavenger
Antimicrobials	
acetic acid	disrupts cell membrane function (bacteria, yeasts, some molds)
benzoic acid	disrupts cell membrane function/inhibits enzymes (molds, yeasts, some bacteria)
natamycin	binds sterol groups in fungal cell membrane (molds, yeasts)
nisin	disrupts cell membrane function (gram-positive bacteria, lactic acid-producing bacteria)
nitrates, nitrites	inhibits enzymes/disrupts cell membrane function (bacteria, primarily Clostridium botulinum)
propionic acid	disrupts cell membrane function (molds, some bacteria)
sorbic acid	disrupts cell membrane function/inhibits enzymes/inhibits bacterial spore germination (yeasts, molds, some bacteria)
sulfites and sulfur dioxide	inhibits enzymes/forms addition compounds (bacteria, yeasts, molds)

Antioxidants

The oxidation of food products involves the addition of an oxygen atom to or the removal of a hydrogen atom from the different chemical molecules found in food. Two principal types of oxidation that contribute to food deterioration are autoxidation of unsaturated fatty acids (i.e., those containing one or more double bonds between the carbon atoms of the hydrocarbon chain) and enzyme-catalyzed oxidation.

The autoxidation of unsaturated fatty acids involves a reaction between the carbon-carbon double bonds and molecular oxygen (O_2). The products of autoxidation, called free radicals, are highly reactive, producing compounds that cause the off-flavours and off-odours characteristic of oxidative rancidity. Antioxidants that react with the free radicals (called free radical scavengers) can slow the rate of autoxidation. These antioxidants include the naturally occurring tocopherols (vitamin E derivatives) and the synthetic compounds butylated hydroxyanisole (BHA), butylated hydroxytoluene (BHT), and tertiary butylhydroquinone (TBHQ).

Specific enzymes may also carry out the oxidation of many food molecules. The products of these oxidation reactions may lead to quality changes in the food. For example, enzymes called phenolases catalyze the oxidation of certain molecules (e.g., the amino acid tyrosine) when fruits and vegetables, such as apples, bananas, and potatoes, are cut or bruised. The product of these oxidation reactions, collectively known as enzymatic browning, is a dark pigment called melanin. Antioxidants that inhibit enzyme-catalyzed oxidation include agents that bind free oxygen (i.e., reducing agents), such as ascorbic acid (vitamin C), and agents that inactivate the enzymes, such as citric acid and sulfites.

Antimicrobials

Antimicrobials are most often used with other preservation techniques, such as refrigeration, in order to inhibit the growth of spoilage and pathogenic microorganisms. Sodium chloride (NaCl), or common salt, is probably the oldest known antimicrobial agent. Organic acids, including acetic, benzoic, propionic, and sorbic acids, are used against microorganisms in products with a low pH. Nitrates and nitrites are used to inhibit the bacterium *Clostridium botulinum* in cured meat products (e.g., ham and bacon). Sulfur dioxide and sulfites are used to control the growth of spoilage microorganisms in dried fruits, fruit juices, and wines. Nisin and natamycin are preservatives produced by microorganisms. Nisin inhibits the growth of some bacteria, while natamycin is active against molds and yeasts.

Sensory Agents

Colorants

Colour is an extremely important sensory characteristic of foods; it directly influences the perception of both the flavour and quality of a product. The processing of food can cause degradation or loss of natural pigments in the raw materials. In addition, some formulated products, such as soft drinks, confections, ice cream, and snack foods, require the addition of colouring agents. Colorants are often necessary to produce a uniform product from raw materials that vary in colour intensity. Colorants used as food additives are classified as natural or synthetic. Natural colorants are derived from plant, animal, and mineral sources, while synthetic colorants are primarily petroleum-based chemical compounds.

Soft drink: Many soft drinks, including colas, contain colorants.

Natural Colorants

Most natural colorants are extracts derived from plant tissues. The use of these extracts in the food industry has certain problems associated with it, including the lack of consistent colour intensities, instability upon exposure to light and heat, variability of supply, reactivity with other food components, and addition of secondary flavours and odours. In addition, many are insoluble in water and therefore must be added with an emulsifier in order to achieve an even distribution throughout the food product.

Natural food colorants				
chemical class	colour	plant source	pigment	products
anthocyanins	red	strawberry (Fragaria species)	pelargonidin 3-glucoside*	beverages, confections, preserves, fruit products
	blue	grape (Vitis species)	malvidin 3-glucoside*	beverages

betacyanins	red	beetroot (Beta vulgaris)	betanin	dairy products, desserts, icings
carotenoids**	yellow/orange	annatto (Bixa orellana)	bixin	dairy products, margarine
	yellow	saffron (Crocus sativus)	crocin	rice dishes, bakery products
	red/orange	paprika (Capsicum annuum)	capsanthin	soups, sauces
	orange	carrot (Daucus carota)	beta-carotene	bakery products, confections
	red	mushroom (Cantharellus cinnabarinus)	canthaxanthin	sauces, soups, dressings
phenolics	orange/yellow	turmeric (Cuycuma longa)	curcumin	dairy products, confections
*Plus other similar compounds.				
**Many carotenoids used as food colorants are chemically synthesized.				

Synthetic Colorants

Synthetic colorants are water-soluble and are available commercially as powders, pastes, granules, or solutions. Special preparations called lakes are formulated by treating the colorants with aluminum hydroxide. They contain approximately 10 to 40 percent of the synthetic dye and are insoluble in water and organic solvents. Lakes are ideal for use in dry and oil-based products. The stability of synthetic colorants is affected by light, heat, pH, and reducing agents. A number of dyes have been chemically synthesized and approved for usage in various countries. These colorants are designated according to special numbering systems specific to individual countries. For example, the United States uses FD&C numbers (chemicals approved for use in foods, drugs, and cosmetics), and the European Union (EU) uses E numbers.

Synthetic food colorants			
	designation		
common name	United States	European Union	products
allura red AC	FD&C red no. 40	...	gelatin, puddings, dairy products, confections, beverages
brilliant blue FCF	FD&C blue no. 1	E133	beverages, confections, icings, syrups, dairy products
erythrosine	FD&C red no. 3	E127	maraschino cherries
fast green FCF	FD&C green no. 3	...	beverages, puddings, ice cream, sherbet, confections
indigo carmine	FD&C blue no. 2	E132	confections, ice cream, bakery products
sunset yellow FCF	FD&C yellow no. 6	E110	bakery products, ice cream, sauces, cereals, beverages
tartrazine	FD&C yellow no. 5	E102	beverages, cereals, bakery products, ice cream, sauces

All synthetic colorants have undergone extensive toxicological analysis. Brilliant Blue FCF, Indigo Carmine, Fast Green FCF, and Erythrosine are poorly absorbed and show little toxicity. Extremely high concentrations (greater than 10 percent) of Allura Red AC cause psychotoxicity, and Tartrazine induces hypersensitive reactions in some persons. Synthetic colorants are not universally approved in all countries.

Flavourings

The flavour of food results from the stimulation of the chemical senses of taste and smell by specific food molecules. Taste reception is carried out in specialized cells located in the taste buds. The five basic taste sensations—sweet, salty, bitter, sour, and umami—are detected in regions of the tongue, mouth, and throat. Taste cells are specific for certain flavour molecules (e.g., sweeteners).

In addition to the basic tastes, the flavouring molecules in food stimulate specific olfactory (smell) cells in the nasal cavity. These cells can detect more than 10,000 different stimuli, thus fine-tuning the flavour sensation of a food.

A flavour additive is a single chemical or blend of chemicals of natural or synthetic origin that provides all or part of the flavour impact of a particular food. These chemicals are added in order to replace flavour lost in processing and to develop new products. Flavourings are the largest group of food additives, with more than 1,200 compounds available for commercial use. Natural flavourings are derived or extracted from plants, spices, herbs, animals, or microbial fermentations. Artificial flavourings are mixtures of synthetic compounds that may be chemically identical to natural flavourings. Artificial flavourings are often used in food products because of the high cost, lack of availability, or insufficient potency of natural flavourings.

Flavour enhancers are compounds that are added to a food in order to supplement or enhance its own natural flavour. The concept of flavour enhancement originated in Asia, where cooks added seaweed to soup stocks in order to provide a richer flavour to certain foods. The flavour-enhancing component of seaweed was identified as the amino acid L-glutamate, and monosodium glutamate (MSG) became the first flavour enhancer to be used commercially. The rich flavour associated with L-glutamate was called umami.

Other compounds that are used as flavour enhancers include the 5′-ribonucleotides, inosine monophosphate (IMP), guanosine monophosphate (GMP), yeast extract, and hydrolyzed vegetable protein. Flavour enhancers may be used in soups, broths, sauces, gravies, flavouring and spice blends, canned and frozen vegetables, and meats.

Sweeteners

Sucrose, or table sugar, is the standard on which the relative sweetness of all other sweeteners is based. Because sucrose provides energy in the form of carbohydrates, it is considered a nutritive sweetener. Other nutritive sweeteners include glucose, fructose, corn syrup, high-fructose corn syrup, and sugar alcohols (e.g., sorbitol, mannitol, and xylitol).

Efforts to chemically synthesize sweeteners began in the late 1800s with the discovery of saccharin. Since then, a number of synthetic compounds have been developed that provide few or no calories or nutrients in the diet and are called nonnutritive sweeteners. These sweeteners have significantly greater sweetening power than sucrose, and therefore a relatively low concentration may be used in food products. In addition to saccharin, the most commonly used nonnutritive sweeteners are cyclamates, aspartame, and acesulfame K.

The sensation of sweetness is transmitted through specific protein molecules, called receptors, located on the surface of specialized taste cells. All sweeteners function by binding to these receptors on the outside of the cells. The increased sweetness of the nonnutritive sweeteners relative to sucrose may be due to either tighter or longer binding of these synthetic compounds to the receptors.

Nonnutritive sweeteners are primarily used for the production of low-calorie products including baked goods, confectioneries, dairy products, desserts, preserves, soft drinks, and tabletop sweeteners. They are also used as a carbohydrate replacement for persons with diabetes mellitus and in chewing gum and candies to minimize the risk of dental caries (i.e., tooth decay). Unlike nutritive sweeteners, nonnutritive sweeteners do not provide viscosity or texture to products, so bulking agents such as polydextrose are often required for manufacture.

Toxicological Testing and Health Concerns

Food additives and their metabolites are subjected to rigorous toxicological analysis prior to their approval for use in the industry. Feeding studies are carried out using animal species (e.g., rats, mice, dogs) in order to determine the possible acute, short-term, and long-term toxic effects of these chemicals. These studies monitor the effects of the compounds on the behaviour, growth, mortality, blood chemistry, organs, reproduction, offspring, and tumour development in the test animals over a 90-day to two-year period. The lowest level of additive producing no toxicological effects is called the no-effect level (NOEL). The NOEL is generally divided by 100 to determine a maximum acceptable daily intake (ADI).

Toxicological analysis of the nonnutritive sweeteners has produced variable results. High concentrations of saccharin and cyclamates in the diets of rats have been shown to induce the development of bladder tumours in the animals. Because of these results, the use of cyclamates has been banned in several countries, including the United States, and the use of saccharin must include a qualifying statement regarding its potential health risks. However, no evidence of human bladder cancer has been reported with the consumption of these sweeteners. Both aspartame and acesulfame K have been deemed to be relatively safe, with no evidence of carcinogenic potential in animal studies.

Canning

Canning is an important, safe method for preserving food if practiced properly. The canning process involves placing foods in jars or similar containers and heating them to a temperature that destroys micro-organisms that cause food to spoil. During this heating process air is driven out of the jar and as it cools a vacuum seal is formed. This vacuum seal prevents air from getting back into the product bringing with it contaminating micro-organisms.

Safe Canning Methods

There are two safe ways of processing food, the boiling water bath method and the pressure canner method:

- The boiling water bath method is safe for tomatoes, fruits, jams, jellies, pickles and other preserves. In this method, jars of food are heated completely covered with boiling water (212° F at sea level) and cooked for a specified amount of time.

- Pressure canning is the only safe method of preserving vegetables, meats, poultry and seafood. Jars of food are placed in 2 to 3 inches of water in a special pressure cooker which is heated to a temperature of at least 240° F. This temperature can only be reached using the pressure method. A microorganism called Clostridium botulinum is the main reason why pressure processing is necessary. Though the bacterial cells are killed at boiling temperatures, they can form spores that can withstand these temperatures. The spores grow well in low acid foods, in the absence of air, such as in canned low acidic foods like meats and vegetables. When the spores begin to grow, they produce the deadly botulinum toxins(poisons).

The only way to destroy these spores is by pressure cooking the food at a temperature of 240° F, or above, for a specified amount of time depending on the type of food and altitude. Foods that are low acid have a pH of more than 4.6 and because of the danger of botulism, they must be prepared in a pressure canner.

The low acidic foods include:

- Meats
- Seafood
- Poultry
- Dairy products
- All vegetables

High acid foods have a pH of 4.6 or less and contain enough acid so that the Clostridium botulinum spores can not grow and produce their deadly toxin. High acidic foods can be safely canned using the boiling water bath method.

The high acidic foods include:

- Fruits
- Properly pickled vegetables

Certain foods like, tomatoes and figs, that have a pH value close to 4.6 need to have acid added to them in order to use the water bath method. This is accomplished by adding lemon juice of citric acid.

Canning Equipment

Water Bath Canners

A water bath canner is a large cooking pot, with a tight fitting lid and a wire or wooden rack that keeps jars from touching each other. The rack allows the boiling water to flow around and underneath jars for a more even processing of the contents. The rack also keeps jars from bumping each other and cracking or breaking. If a rack is not available, clean cotton dish towels or similar can be used to pack around jars. If a standard canner is not available any large metal container may be used as long as it is deep enough for l to 2 inches of briskly boiling water to cover the jars. The diameter of the canner should be no more than 4 inches wider than the diameter of your stove's burner to ensure proper heating of all jars. Using a wash kettle that fits over two burners is not recommended because the middle jars do not get enough heat. For an electric range, the canner must have a flat bottom. Outdoor fire pits with a solid grate will also work however close attention is required to insure proper boiling temperature.

Pressure Canners

A pressure canner is a specially-made heavy pot with a lid that can be closed steam-tight. The lid is fitted with a vent (or pet-cock), a dial or weighted pressure gauge and a safety fuse. Newer models have an extra cover-lock as an added precaution. It may or may not have a gasket. The pressure pot also has a rack. Because each type is different, be sure to read the directions for operating.

Jars

Mason jars and Ball jars specifically designed for home canning are best. Commercial mayonnaise jars, baby food and pickle jars should not be used. The mouths of the jars may not be appropriate for the sealing lids and the jars are not made with heavy glass and they are not heat treated.

Jars come in a variety of sizes from half-pint jars to half-gallon jars. Pint and quart Ball jars are the most commonly used sizes and are available in regular and wide-mouth tops. If properly used, jars may be reused indefinitely as long as they are kept in good condition.

Atlas jars should not be used for home preserving and canning.

Jar Lids

Most canning jars sold today use a two piece self-sealing lid which consists of a flat metal disc with a rubber-type sealing compound around one side near the outer edge, and a separate screw-type metal band. The flat lid may only be used once but the screw band can be used over as long as it is cleaned well and does not begin to rust.

Canning Utensils

Helpful items for home canning and preserving:

- Jar lifter: essential for easy removal of hot jars.
- Jar funnel: helps in pouring and packing of liquid and small food items into canning jars.

- Lid wand: magnetized wand for removing treated jar lids from hot water.
- Clean cloths: handy to have for wiping jar rims, spills and general cleanup.
- Knives: for preparing food.
- Narrow, flat rubber spatula: for removing trapped air bubbles before sealing jars.
- Timer or clock: for accurate food processing time.
- Hot pads.
- Cutting board.

There are also many specialty utensils available like apple slicers, cutting spoons for coring and pit removal, corn cutters and fruit skinners.

Food Fortification

Fortification is the addition of key vitamins and minerals such as iron, iodine, zinc, Vitamin A & D to staple foods such as rice, milk and salt to improve their nutritional content. These nutrients may or may not have been originally present in the food before processing.

70% of people in India do not consume enough micronutrients such as vitamins and minerals. About 70 percent of pre-school children suffer from anaemia caused by Iron Deficiency and 57 percent of preschool children have sub–clinical Vitamin A deficiency. Neural Tube Defects (NTDs) are the most common congenital malformation with an incidence that varies between 0.5-8/1000 births. It is estimated that 50-70% of these birth defects are preventable. One of the major causes is deficiency of Folic Acid.

Thus, deficiency of micronutrients or micronutrient malnutrition, also known as "hidden hunger", is a serious health risk. Unfortunately, those who are economically disadvantaged do not have access to safe and nutritious food. Others either do not consume a balanced diet or lack variety in the diet because of which they do not get adequate micronutrients. Often, there is considerable loss of nutrients during the processing of food. One of the strategies to address this problem is fortification of food. This method complements other ways to improve nutrition such as such as diversification of diet and supplementation of food.

Benefits of Fortification

- Since the nutrients are added to staple foods that are widely consumed, this is an excellent method to improve the health of a large section of the population, all at once.
- Fortification is a safe method of improving nutrition among people. The addition of micro-nutrients to food does not pose a health risk to people. The quantity added is so small and so well regulated as per prescribed standards that likelihood of an overdose of nutrients is unlikely.

- It does not require any changes in food habits and patterns of people. It is a socio-culturally acceptable way to deliver nutrients to people.
- It does not alter the characteristics of the food—the taste, the feel, the look.
- It can be implemented quickly as well as show results in improvement of health in a relatively short period of time.
- This method is cost-effective especially if advantage is taken of the existing technology and delivery platforms.
- The Copenhagen Consensus estimates that every 1 Rupee spent on fortification results in 9 Rupees in benefits to the economy. It requires an initial investment to purchase both the equipment and the vitamin and mineral premix, but overall costs of fortification are extremely low. Even when all program costs are passed on to consumers, the price increase is approximately 1-2%, less than normal price variation. Thus it has a high benefit-to-cost ratio.

Vacuum Packaging

Vacuum packaging is an easy and efficient way of packaging a variety of products. Vacuum packaging is often used in commercial and residential applications to preserve food. Technically it is considered a form of modified atmosphere packaging (MAP). This is any form of packaging the atmosphere around the product is altered from the external atmosphere.

Before pointing out the advantages, we will describe the start-up machines and materials necessary to begin vacuum packaging.

Vacuum sealers can vary in price and requirements often based on production levels the machine sizes. Determining projected production levels and product sizes prior to purchasing a machine can save on frustration down the road. Another point of consideration should be the option of gas flush. Gas flush injects an inert gas into the sealed package to help increase shelf life.

Once the proper sealer is determined the vacuum bags should be chosen. Vacuum bags come with in a variety of sizes with multiple options available. Determining the right vacuum bag for each product will help ensure sleek and professional packaging.

Vacuum Packaging Benefits

- Increased Shelf Life: Studies have shown that vacuum packaging can increase product shelf life from 50%-400%. The key to increasing product shelf life is determining the ideal atmosphere within the packaging for the product being packaged. This could be the introduction of an inert gas or the reduction of oxygen to ideal levels. Vacuum packaging is also and preferred form of packaging for many products going into long term freezer storage. The proper vacuum bags can dramatically reduce freezer burn. It is a chosen form of packaging for meat storage around the world.

- Reduced Product Loss: Increased shelf life directly effects the reduction in product loss for companies inventorying products. The reduced product loss helps to increase bottom lines.
- Sealed Barrier From External Elements: The most common vacuum bag thickness is a 3 mil thickness with some bags up to a 6-8 mil thickness. With the proper seal, vacuum packaging can protect products from dust, moisture, insects, and a variety of other external elements that may harm or damage products. Vacuum packaging is also an excellent barrier of protection against freezer burn during long term freezer storage.

- Seals in Flavors**:** Wet aging meats is a popular form of curing meat before preparing it. This often uses natural juices from the meat with spices and flavors vacuum sealed to lock in flavors during the curing process. Check out this video on wet aging steaks for more information.
- No Chemical Preservatives Required: Vacuum packaging eliminates the need for chemical preservatives. With the right mixture of oxygen to inert gas, products can last longer without the use of commonly used preservatives. The earth's atmosphere is made up of 78% nitrogen which is the most commonly used inert gas for vacuum packaging.
- Improved Product Presentation: Most vacuum bags are crystal clear allowing products within the packaging to be displayed on shelves. Vacuum packaging is an affordable packaging solution the enables products to be shelf ready with the application of a custom printed sticker.

- Multiple Packaging Options: Vacuum sealing materials come with multiple packaging options. Users can buy rolls of film or bags to seal products. Bags can be found with reclosable zippers, easy open tear notches, foil lined, channeled, and with a variety of other options to choose from.
- Quick and Efficient Packaging: With the proper machines and materials, vacuum sealing can be a very efficient process enabling hundreds or thousands of products to be sealed per hour.

Polyethylene Bags

Plastic bags are made from crude oil, which is made into a hard or a soft material. Plastic bags have only existed for one generation. In this short time, their extreme convenience has made them widespread all aspects of life, making them the synonym of waste. However, their decomposition is a very long process; in fact, they never fully decompose in seawater.

Plastic bags can be classified into 3 main groups according to decomposition:

- Polyethylene plastic bags

These are the most common plastic bags that are made from crude oil. Their decomposition takes 700 – 1000 years, and in unfavourable conditions they never decompose.

Plastic bags decompose into smaller pieces, emitting toxic compounds in the process. They also make deadly decoy food for fish, birds, and animals. For example, plastic bags have clogged the sewage drains in Bangladesh, resulting in a flood that killed thousands of people.

Although polyethylene plastic can be recycled, only an estimated 1 – 3% of this reaches the recycling process.

- Plastic bags with an accelerated decomposition process

These plastic bags are often mistakenly called biodegradable plastic bags. In addition to polyethylene, these plastic bags contain molecules of vegetable origin. In favourable conditions, the plastic is colonized by bacteria. They eat the filler and thus, make several holes in the plastic, resulting in its decreased porosity, which, in turn, contributes to its further decomposition in various ways, including by bacteria.

There have been many disputes about the environmental sustainability of this type of plastic bags, because although the decomposition process is accelerated, the bags are sometimes never really destroyed. They only disintegrate into smaller pieces that are a burden for the environment and a threat to birds and animals. Through the food chain, plastic even makes its way onto our table. The organism of a modern human being contains about 100 synthetic chemical compounds that were not there 50 years ago.

This type of plastic bags is definitely unsuitable for organic waste containers and also not really appropriate for plastic waste containers, because they impair the entire recycling process.

- 100% biodegradable plastic bags

These plastic bags are made of 100% bacterial degradable material, such as corn, coconut, or rice starch, soya protein, etc. However, this type of plastic bags is less resistant to wear and tear and also more expensive compared to polyethylene plastic bags. Nonetheless, they are 100% biodegradable in favourable conditions and the decomposition process only takes a short amount of time. However, their production costs increase the world market prices of food products. These plastic bags can be cumulated in compost heaps.

Uses and Application of Poly Bags

The uses for poly bags are virtually without limit! Whenever there is a need to protect an item during storage or shipment, Material Concepts can supply a poly bag that is right for the job. Lighter density poly bags are a good choice for items that are not heavy and require only a simple shield from their surroundings. Heavy density poly bags keep goods safe from harm in more abusive environments where dirt, dust, moisture, movement, etc. are factors that may damage goods.

The shape of the bag is also an important consideration in some cases. Flat poly bags are general purpose bags that are often used where a clear presentation at a low cost is desired. Material Concepts offers wicketted poly bags and gusseted poly bags for applications where the shape and style of the poly bag is vital to how it will be used. For protection against moisture in particular, a reclosable poly bag is suggested, keeping goods safe while allowing for easy opening and closing.

Garment poly bags are specially designed and shaped for use with shirts, coats, pants, etc. for protection and ease of use in apparel applications. Poly film and poly tubing are used where large quantities are required and use of a roll of the film or tubing is a good solution.

Wooden Field Boxes

A wooden box is a container made of wood for storage or as a shipping container. Wooden Boxes have been used from several years for storage needs. Wooden Boxes as a storage means are a best way to give protection for the objects that are being stored as these boxes can't easily be damaged or broken. Wooden boxes are often used for heavy duty packaging when high strength is needed for heavy and difficult loads and also when long term warehousing is needed. Wooden Boxes have a bunch of benefits as a storage material. Some storage benefits of Wooden Boxes are as follows:

- Wooden Boxes can be used for years as they are not easily breakable and can be used for longer times for storing different items.
- Wooden Boxes are used widely by shipping industries for safe delivery of imported and exported items.
- Wooden Boxes are good to protect the thing kept inside as it don't allow water to go inside, making the storage item safe inside the box.
- Wooden Boxes are essential to store essential oils, so as to protect them from losing essence.

References

- Sanitizers-and-disinfectants-the-chemicals-of-prevention: foodsafetymagazine.com, Retrieved 20 April 2018
- Fruit-waxing, dietandnutrition: medindia.net, Retrieved 22 March 2018
- What-is-the-importance-of-food-preservation-201105176772: preservearticles.com, Retrieved 26 May 2018
- Food-additive: britannica.com, Retrieved 09 July 2018
- Fortification-of-food, nutrition: vikaspedia.in, Retrieved 29 June 2018
- 8-benefits-of-vacuum-packaging: uspackagingandwrapping.com, Retrieved 31 March 2018

5

Machines used for Postharvest Handling

The most important goal of post-harvest handling is preservation of the agricultural produce by keeping the product cool, avoiding moisture loss and delaying any unwanted chemical change. Sanitation is another consideration. The diverse machinery that helps to ensure such objectives have been extensively covered in this chapter. These include cartooning machines, form, fill and seal machines, grain dryer, rice huller, hay rake, etc.

Cartoning Machines

A cartoning machine, sometimes called a cartoner for short, is a type of packaging machine. Its sole purpose is to form cartons. Have you ever seen a milk carton? If you look at the average milk carton in the grocery store, a cartooning machine probably made it. A cartooning machine forms cartons that stand up straight, close, are folded, side-seamed, and then, ultimately, sealed.

Cartoning machines can be sub-divided into two basic types:

- Vertical cartooning machines
- Horizontal cartooning machines

A carton machine will pick up a single piece from a stack of fold carton and then get it to stand up straight, or erect. The machine will fill it with a product or a number of products horizontally through an end that is open, and then close it by tucking an end flap of the carton or putting on glue or adhesive. It is not that hard to understand, but for those who aren't involved in the cartoning field, it can be a little tricky to understand if they haven't seen one first-hand in action. The product can be pushed into the carton with either pressurized air or with a mechanical sleeve. Technology is always changing, and newer, better, and more economical ways of doing the same job are always coming out. However, for a number of applications, the products are inserted into the carton by hand. A cartoning machine is often used for packaging sundry goods, cosmetics, confectionary, foodstuffs, etc.

A cartoning machine which produces a folded carton, fills it up with a product or several products vertically through an end that is open, and then closes it by tucking in the end flap and applying glue or adhesive, is called an end-load vertical cartoning machine. Cartoning machines are used on a regular basis for packaging medicine, confectionary, cosmetics, etc.

Checkweigher

A checkweigher is usually part of a typical quality control system, ensuring that every product leaving a production line is the right weight and corresponds to packaging requirements.

Checkweighers are key to ensuring manufacturers deliver quality products that ensure customer satisfaction. Selling products with incorrect weights causes problems for manufacturers: underweight products can result in companies being fined or subject to other penalties, whilst overweight products generate unnecessary and expensive product giveaway. Checkweighers can be used in many industries as part of an effective quality assurance system. Relevant sectors include:

- Food
- Pharmaceuticals
- Cosmetics
- Beverages
- Transport/logistics
- Chemicals
- Automotive
- Metal manufacture and fabrication industries

In an increasingly competitive marketplace, checkweighers are essential in fulfilling ever-changing customer needs and in complying with local Weights & Measures standards, as well as global standards.

Components of a Checkweigher

There are several different checkweigher designs in common use, with components varying greatly depending on: how they are used, items being weighed, and their surrounding environment.

A typical checkweighing system comprises:

- An in feed section
- A weighing section
- An out feed section with sorting/rejecting device
- A weighing terminal as user interface

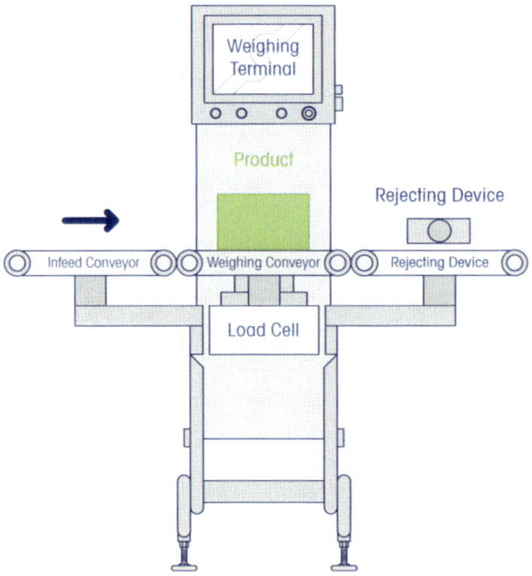

(Checkweighers are also available with metal detection, X-ray inspection, vision inspection combinations)

Checkweighers are used as part of overall quality control programmes, in which manufacturers can be assured that they are complying with regulations and are supplying products in the right amounts. In addition, consumers can be assured that they are acquiring high-quality products that contain all the ingredients or parts, as specified on the label.

Difference between a Static Scale and a Dynamic Checkweigher

The main differences between static scales and dynamic checkweighers are:

- Static scales measure the weight of products that stand still, whereas. Dynamic checkweighers weigh products in motion.
- Static scales are used for manual weighing of products or for sample spot-checking, whereas Dynamic checkweighers automatically check 100 % of the products produced.
- In order to check product weights, static scale weighing is required. This is a manual operation requiring product weighing, recording the result, and then product removal for the next weighing procedure, whereas Dynamic checkweighing is entirely automatic, without manual intervention or a dedicated operator: products are weighed as they pass along a production line, with off-weight products rejected from the production line by automated removal devices such as pneumatic pusher arms or air-jets (sometimes known as 'air-blasts').

In a production line, static scales are used to sample a certain percentage of product weights, whilst dynamic checkweighers automatically sample 100 % of all weights.

Both static scales and dynamic checkweighers are to be found in most production environments, with static scales commonly:

- Determining target weights for dynamic checkweighers.
- Performing sample tests for net weight and tare weight reports that satisfy Weights & Measures compliance requirements and Weights & Measures package tare weight verification reports.

The choice of checkweighing system used on a production line is usually dictated by application requirements and processes, financial and economic factors, plus the need for brand protection.

Typical Uses

Checkweighers are used for a variety of applications some of which include:

- Checking for under and overweight packages.
- Ensuring compliance with net contents laws for pre-packaged goods.
- Checking for missing components in a package including instructions, lids, leaflets or products.
- Verifying count by weight by checking for a missing carton, bottle, bag or can in a case.
- Checking package mixes against weight limits to keep the solid to liquid ratio within established standards.

- Reducing product giveaway by using checkweigher values to determine filler adjustments.
- Classifying products into weight zones for grading or portioning.
- Ensuring product compliance with customer, association or agency specifications.
- Pure 'net weighing' with tare/gross systems.
- Weighing before and after a process to check process performance.
- Fulfilling reporting standards in line with USDA, FDA, OIML, FPVO and others.
- Measuring and reporting production line efficiency.

By definition, a checkweigher is a system that weighs items as they pass through a production line, then classifies the items by preset weight zones, and sorts (or rejects) items based on those classifi ations. Checkweighers weigh 100 % of the items on a production line and provide a 100 % overview of production data, such as production counts, batch tracking, total weights, good weights and rejected weights.

Reasons, Advantages and Benefits

Avoid Costly Fines and Tarnished Brand with 100 % Weight Inspection

One of the main reasons for using a checkweigher is to ensure compliance with government regulations and industry standards. 100 % weight inspection should be an integral part of a co-ordinated quality and process control programme – and use of a checkweigher also means that information previously collected manually can now be automatically collected in a fraction of the time.

Example: Consider a line with a throughput of 100 packages per minute. If you sample 15 packages every hour, what percentage of the total production is that sample? In 60 minutes, 60 x 100 = 6000 packages pass through the line. 15 packages represent only 15/6000 = 0.25 %. With a sample size that small, there is a little statistical signifi ance to the results, because over 99.75 % of packages are not inspected. A checkweigher automatically weighs 100 % of all packages on the line and can react immediately if it senses a problem.

Protection of the Customer and Consumer

To ensure customer and end-consumer satisfaction, manufacturers have an obligation to ensure consistent product quality by minimising the risk of producing under- filled products or supplying products with missing parts. Failure to achieve these aims can create potential animosity between the retailer or customer and the manufacturer, and can result in the potential breakdown of customer relationships plus loss of future business opportunities.

Net content laws and regulations differ from country to country; for example, in the US the National Institute of Standards and Technology's (NIST1) Handbook 133 on pre-packaged goods defines the specifi net contents laws on packages for processors, wholesalers and retailers. It states that the average weight of the product cannot be less than the label weight. It also states that no individual package can exceed the "Maximum Allowable Variation (MAV)" which is the amount of allowed underweight for a certain labelled weight.

As another example, the OIML22 R87 (International Organisation of Legal Metrology) standard is used by most European nations and many other countries around the world. The standard defines the tolerable deficiencies in actual content for pre-packaged products, whilst national Weights & Measures authorities regulate and enforce the local net content laws applicable in different countries. Official action resulting from package checking can take the form of spoken recommendations, instructions, warnings, or legal action.

Protection of Brand and Reputation

Powerful and positive product branding gives customers a strong assurance of safety and quality. Branding is frequently responsible for driving consumer repeat purchases, and so is an important tool in maximising sales and justifying premium product pricing by manufacturers and retailers.

For these reasons, a company's responsibility is not only related to protection of the end consumer but also to the brand and reputation of the company. Product brands are important assets that need to be managed carefully and protected from adverse publicity. Underweight product acquired by consumers can have a negative impact on an organisation, resulting in damage to the brand and potentially costly recalls.

In the event of a company being investigated as a result of a customer complaint, documentation will provide invaluable evidence of the correct operation of the checkweighing programme.

Counting and Statistical Functions

Checkweighers provide real-time monitoring of production processes, including yield statistics and SPC (Statistical Process Control) trending. This can be used for process improvements and operating efficiencies.

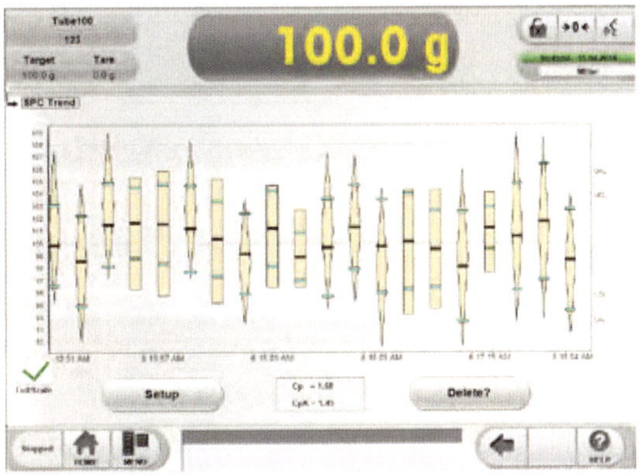

Figure: Real-time monitoring of production processes

Examples of checkweigher statistical functions and production line efficiency functions include:

- Analysing production by weight zone or classification.
- Using 3 or more zones to gain detailed fill weight information.
- Monitoring overall production efficiency through total count and total weight.

- Monitoring overall production rate efficiency (packages per minute).
- Monitoring mean value (average) to alert operators or fillers of an out-of-tolerance condition or trend.

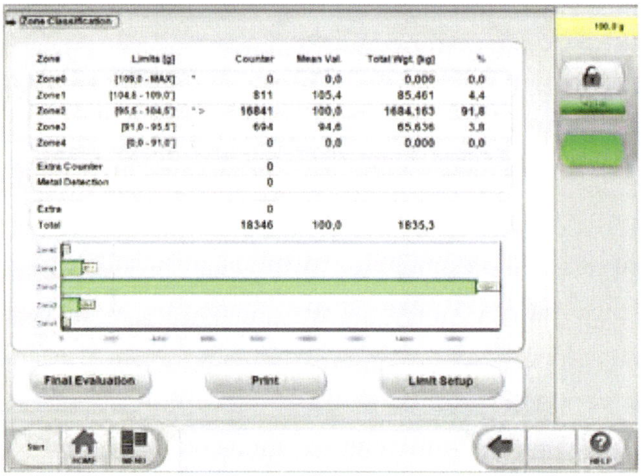

Figure: Monitoring production data

- Maintaining records of settings for management and regulatory agencies.
- Analysing filler head performance for both single and multi-head fillers.
- Printing or accumulating individual production weights or totals for a day, shift, hour, batch or collecting and printing a production run.
- Monitoring short, long-term or individual weights and filler performance through statistics.
- Providing SPC charts for manual feedback and process adjustments.
- Providing closed loop control and automatic process.
- adjustments.
- Linking packaging line data to upstream control and information systems.
- Interfacing with business systems, Programmable Logic Controllers (PLC) and SCADA systems, to link the checkweigher to the production process (including controlling the checkweigher by a remote device).
- Reducing quality control labour.
- Providing an important information source for Quality Control departments.

Making Better use of Limited Resources

An accurate checkweigher will ensure tighter production tolerances that will yield less waste and enable more final products to be produced from the same amount of inventory. Raw materials can be very expensive, and a checkweigher should be an integral part of an overall programme to make the most of existing resources.

Package and Production Information	Savings Yielded (By a 1 Gram Overfill Reduction)
Labelled Package Weight = 450 grams	0.1 ct Savings per Package
Material Cost = 0.1 ct per gram	20 ct Savings per Minute
Line Rate = 200ppm Line Utilisation = 65 % Shift= 8 hours	€12 Savings per Hour
Shifts per Day=2	€124 Savings per Day
Production Days per Year = 230	€28,704 Savings per Year
The reduction of 1 gram overfill as shown in this example would free enough raw materials to produce an additional 60,000 products	

Figure: Shows a simple calculation that demonstrates savings which can be made by reducing overfill by 1 gram

Keep Fillers in Tune with Active Checkweigher Feedback

Feedback to filler is one of the strongest reasons to invest in a checkweigher. It enables active monitoring of filler performance, plus it minimises unwanted overfills and underfills by keeping filler heads and processes properly adjusted. Checkweighers can communicate directly with the filler control; alternatively, they can network to existing factory floor automation systems. The overall result is complete and seamless feedback control.

Fewer False Rejects, Less Rework and Reduced Scrap

Processes that incorporate an accurate and well-maintained checkweigher also find fewer products in their reject bins. Tighter tolerances lead to improved overall process quality, reducing the amount of scrapped product and those requiring re-work. False rejects are also minimised, because checkweigher accuracies are more precise as zone settings become more refined.

Labour-Saving Operation

Use of static scales to spot-check products is common to many operations; however, use of in-line checkweighing eliminates potential sampling errors and also removes often excessive labour costs required for static scale spot-checking.

Increased Overall Equipment Effectiveness

Overall Equipment Effectiveness (OEE) is enhanced through improvement of any or all of OEE's three metrics: availability, performance, and quality. 100 % monitoring of items provides valuable indications of:

- Process variations which can which lead to unexpected downtime events (availability).
- Reduced line capacity (performance) caused by out-of-tune upstream devices.
- Tighter production tolerances arising from checkweighing (reducing overfills and underfills), resulting in a higher yield of satisfactory products (quality).

Inspection of Product Integrity and Package Quality

Checkweighers are also used for additional inspection purposes, above and beyond those described above; in fact checkweighers are, increasingly, becoming quality assurance devices, whilst also integrating other automated inspection processes. Together, these can check factors such as:

- Open flaps on a carton or case.
- Missing caps.
- Bar code labels and tamper evident seals.
- Package orientation and skew detection.
- Printed information on the package, such as batch number, expiry or 'Best Before' date.
- Contaminants such as metal, stone or glass.
- Traceability and serialisation.

Integration of other inspection devices (such as cameras, scanners, marking systems, sensors, metal detectors and X-ray devices) makes checkweighers a high-performance product inspection solution. The main benefit of combining other inspection devices is the consolidation of multiple-user interfaces into one, so as to simplify operator training and reduce time taken for package set-up and line changeovers.

Form, Fill and Seal Machines

Form-fill-seal machines are machines that form the package, fill it with a wet or dry product and seal it closed. Most FFS systems use flexible film to form the primary package, such as a bag or pouch. But gable-top and aseptic cartons are also created in a form-fill-seal operation. And most blister packs are produced with a thermoform/fill/seal machine, a close cousin.

Types of Form, Fill and Seal Machines

Flow-wrapping Machine

Horizontally operating form, fill and seal machine with material reel mounted above the operating level, the product loaded horizontally and a longitudinal seal formed below the pack.

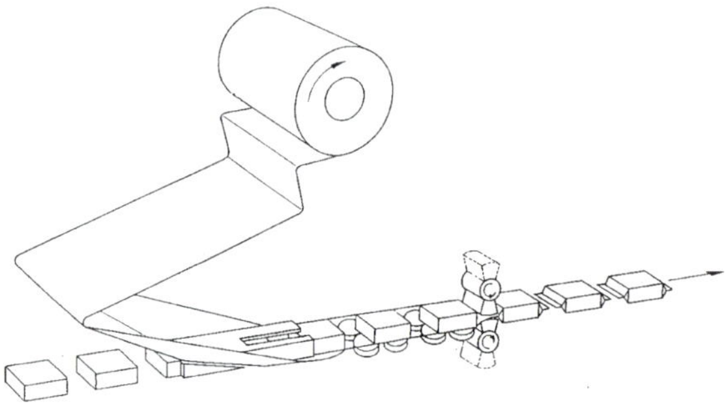

Figure: Diagram showing the principle of operation of a typical flow-wrapping machine

Vertical form, Fill and Seal Machine

Machine which uses flexible packaging material to form a tube, which is then filled vertically with product and sealed in a sequence of operations whilst the packaging material is transported vertically downwards.

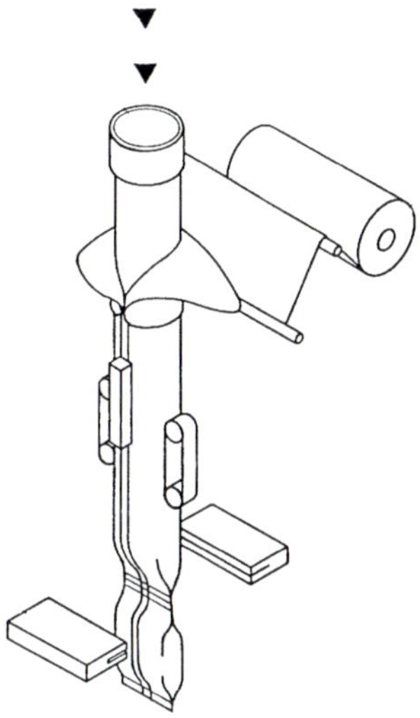

Figure: Diagram showing the principle of operation of a typical vertical form, fill and seal machine

Horizontal Sachet form, Fill and Seal Machine

Horizontally operating form, fill and seal machine in which packs are formed, sealed on 2 or 3 sides, filled vertically with product and sealed on the remaining side whilst the film web is moved horizontally with the pack vertical.

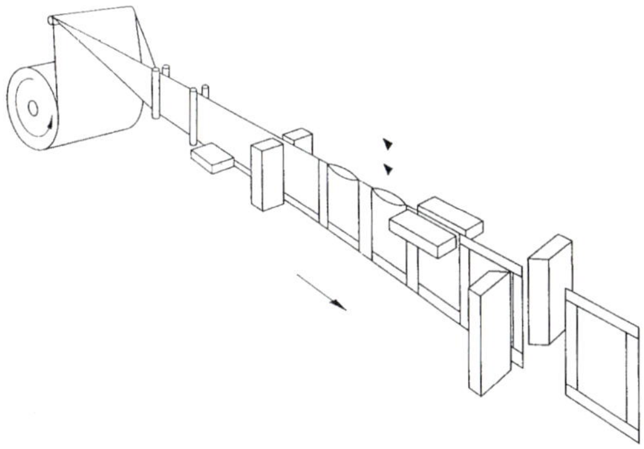

Figure:Diagram showing the principle of operation of a typical horizontal sachet form, fill and seal machine

Thermoform, Fill and Seal Machine

Machine in which a web of thermoformable material is heated and formed with pressure and vacuum, before being filled vertically with product, sealed with a top film or magazine fed lid and finally cut to produce individual packs.

Machines can produce one or more lanes of packs and may incorporate equipment to evacuate packages before they are sealed.

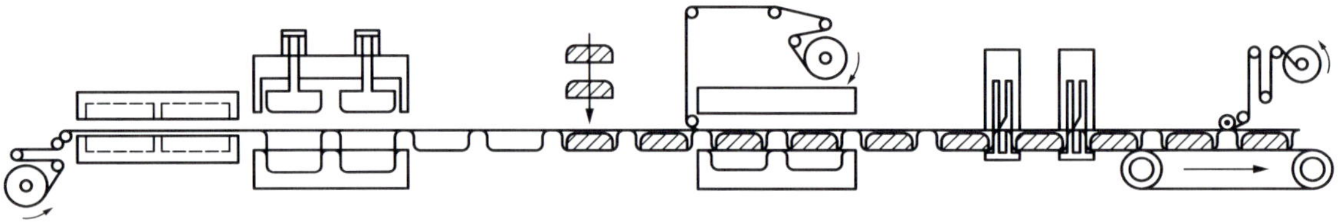

Figure: Diagram showing the principle of operation of a typical thermoform, fill and seal machine

Tubular Bag form, Fill and Seal Machine

Machine which forms a bag or sack from a reel of lay flat tubular flexible packaging film, before it is filled with product and sealed

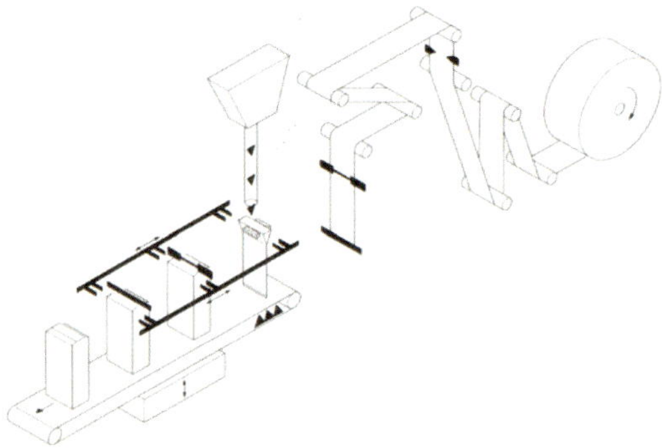

Figure: Diagram showing the principle of operation of a typical tubular bag form, fill and seal machine

Mandrel form, Fill and Seal Machine

Machine which forms packs from a reel of flexible material or carton blanks, on one or more mandrels, before filling the packs with product and sealing their tops within the machine

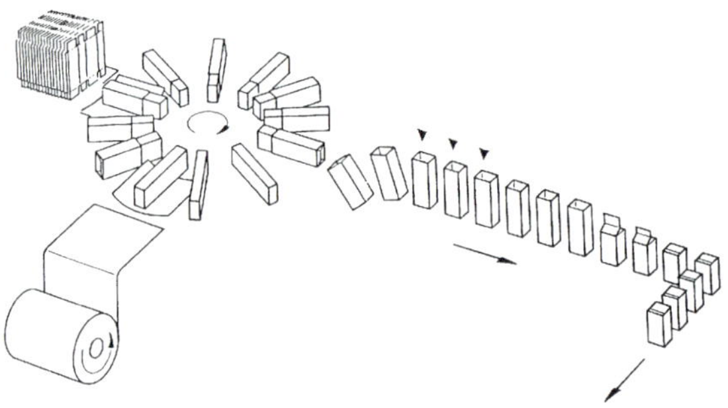

Figure: Diagram showing the principle of operation of a typical mandrel form, fill and seal machine

Types of Fill and Seal Machines

Pre-made Bag Fill and Seal Machine

Machine in which a pre-made bag is taken from a magazine, opened, filled with product and then sealed.

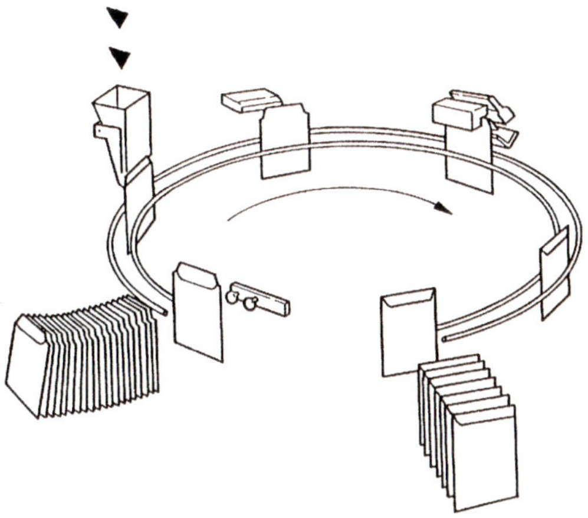

Figure: Diagram showing the principle of operation of a typical pre-made bag fill and seal machine

Sack Fill and Seal Machine

Machine in which a pre-made sack is taken from a magazine, before being opened, filled with product and sealed.

The sacks may be filled in-line or on a rotary carousel.

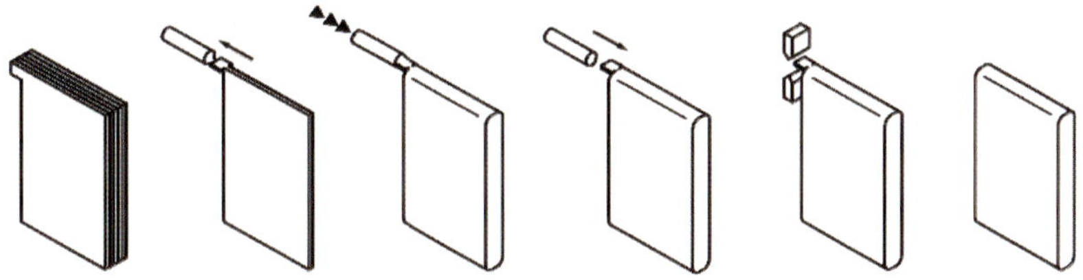

Figure: Diagram showing the principle of operation of a typical in-line sack fill and seal machine

Cup or Tub Fill and Seal Machine

Machine in which a pre-made cup or tub is taken from a magazine, filled and then closed with a heat sealed foil or a press-on lid.

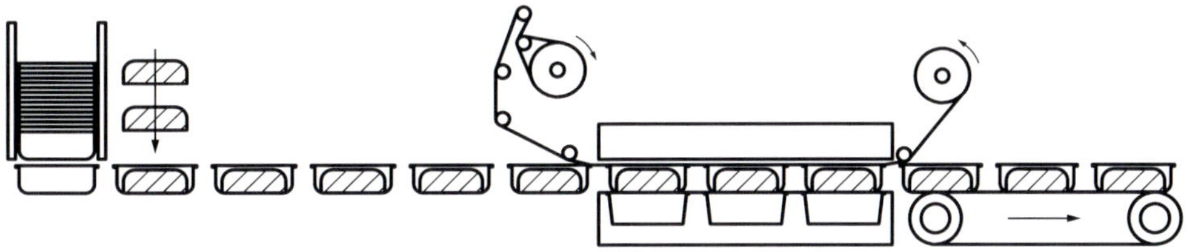

Figure: Diagram showing the principle of operation of a typical cup/tub fill and seal machine

Filling Machines

Auger Filling Machine

Filling machine which measures out a product, usually a powder, using an auger which rotates for a predetermined number of revolutions.

Volumetric Cup Filling Machine

Filling machine which measures out a product, usually free-flowing solids or powder, in a cup of predetermined volume.

Nett Weigher

Filling machine which measures out a predetermined mass of product, usually free-flowing solids, before dispensing it as a fill.

Multi-Head Weigher

Nett weighing machine with multiple weighing units, which computes an appropriate combination of loads to achieve the predetermined mass and discharges them together as a fill.

Autocoding

OAL Autocoding with Cognex

Autocoding refers to software solutions that help manufacturers, particularly those in the food industry, ensure that products have the correct packaging and correct 'sell by' date codes, thereby reducing the number of Emergency Product Withdrawals (EPW). The term was first used during an initiative between Geest PLC and Tesco PLC in 2001.

The key objective of the software was to reduce the number of EPW's associated with date and price coding errors, and pot and lid marriage errors. This still remains the main objective of autocoding software, but functionality has been expanded to encompass quality assurance and OEE performance data.

System Elements

All autocoding systems comprise a products database which contains standard reference information for each product including packaging type, labels and sell by date criteria. In most cases a touch screen industrial PC is positioned on the shop-floor to allow the operator to select the next product from a product schedule.

1D and 2D Barcode Scanning

The shop-floor touch screen device is linked to barcode scanners deployed to scan the code on each piece of packaging, including promotional labels and sleeves. Originally the bar codes scanned were based on standard 1D codes but to avoid mistakes 2D bar codes were introduced in 2004 so that each packaging type could hold a unique identity. To checks that the scanners are operational Autocoding solutions include two way communications with all hardware devices, or prevent the lines starting if links are not available.

Date Code Printing

To ensure that 'sell by dates' are accurate, most autocoding systems directly control the line printers through the software application. Once the operator has selected the product to run, the product reference table identifies the date range to use and the printer output is sent directly to the printer. Again, like the bar code scanners, autocoding systems include two-way communications with date code printers, and prevent the production lines from starting if links are not available.

Line Stops

OAL Autocoding system line stops

If any error is detected, such as wrong film/pack, wrong lid, wrong case, or printer fault, the line is stopped. This is a standard requirement and is achieved through the use of PLCs.

System Providers

Numerous companies provide autocoding solutions, ranging from standalone systems to comprehensive MES/MIS solutions that incorporate additional features and benefits. Notable industrial vendors include:

Product Name	Vendor
Autocoding	OAL Group
AutoCoding	AutoCoding Systems
Harford Control Compliance	Harford Control Ltd
Sparc Systems Ltd	Sparc Systems

Grain Cart

Combine unloading into chaser bin

Chaser bin unloading into a grain hopper trailer.

Two combines unloading into 2 chaser bins

A chaser bin, also called grain cart or (grain) auger wagon, is a trailer towed by a tractor with a built-in auger conveyor system, usually with a large capacity (from several hundred to over 1000 bushels; around 15 tonnes (33,000 lb) is average).

Design

The typical setup of a chaser bin is a cross auger, which feeds the folding unload auger, which in turn empties the contents into waiting Grain hopper trailers or mother bins.

The bins range in size, from 12T to 38T in most cases. Single axle bins usually can only handle a maximum of 20T of grain, with excess weight placing undue strain on the axle, potentially causing failure, especially on rough terrain. Bins above a capacity of 20T, tend to feature a "walking beam" chassis. This chassis design features an independent axle setup, allowing the bin to smoothly travel across ditches and divets, as well as more easily access fields where the road is higher than the field level. It also allows the load to be evenly distributed along the length of the chassis, further reducing the risk of axle failure when traveling over rough terrain, particularly with a full bin.

The design of the bins usually follows the same pattern regardless of capacity, with smooth flowing curves to allow the grain to easily unload via the cross auger. This also helps prevent the grain sticking to the walls of the bin, preventing corrosion.

The cross auger is smaller than the main unload auger, which allows the unload auger to expel the grain at a constant rate. The bins can also feature cross auger cut-offs, which allow the operator to choke the feed rate to the unload auger if it starts struggling. Unload augers can empty a full bin in a matter of minutes, with 15" augers unloading at a rate of 6T per minute, and 19" augers unloading at 10T per minute.

Usage

Chaser bins are typically used to transport harvested grain or corn over fields from a header to a road train or other hauling device which is used to cover larger distances over roads. The use of a chaser bin allows the harvester to operate continuously, eliminating the need to stop and unload.

These bins usually require tractors with large power outputs and are popular on the generally larger and more open fields of the United States and Australia, though usage in Europe is increasing.

Mother bins

Mother bin as a static collection point

Mother bins, also known as "field bins" in Queensland, offer the farmer a convenient storage location while harvesting. They offer a large storage capacity ranging from 55T to 130T. They are usually located on a paddock road at the entrance to the paddock being harvested, and allowing chaser bins to unload into them and offering truck drivers a more convenient location to fill grain into their semi-trailers. This also assists in reducing unnecessary soil compaction, by reducing the number of road trains on a given paddock.

Although they both hold and unload grain in the same manner, mother bins, are not designed to be used as chaser bins. They are only to be moved when empty, or if needed at a maximum of 25% full on smooth terrain only. They are designed to be a static collection point only.

Grain Dryer

A grain dryer, as its name would suggest, is a machine manufactured simply to dry grain. In the United States, this is commonly used for two main crops: corn and soybeans. The grain dryer is an expensive piece of equipment and not one many farmers have. Rather, they may use the services of farming cooperative or rent one in some situations.

Most farmers try to avoid having to use a grain dryer, or at least try to use it very little, whenever possible. This is because of the added expense the additional step will involve. This is why farmers will leave the crops out in the field as long as they possibly dare in the fall. The dead plants will naturally dry out the corn and beans. However, harvest must be done before the plants start dropping the valuable crop.

However, a grain dryer can also be a valuable tool for large-scale farming operations. This is because it allows for earlier harvesting of at least some of the crop. Without advancements in grain dryer technology, that prolonged harvesting season, created by starting earlier, may not have been possible. In the long run, this can be a financial benefit to the farming operation.

A grain dryer is needed simply because of the moisture content inherent in both corn and soybeans in most cases. These crops, if not turned into dry grain, would rot or mold. Either way, the effects for both humans and animals who consumed such grain could be very detrimental, up to and including being fatal. Therefore, using a grain dryer becomes a food safety issue.

The crop is loaded into the grain dryer usually with the help of a grain auger. Once in the dryer, it is subjected to a high level of wind produced by fans and heat, produced by powerful heaters. The temperature of an activated dryer can approach the boiling point. The more air and heat passing through the dryer, the faster the grain will be dried. After that, the grain may be stored in a grain bin or grain silo until ready for further transport.

In some cases, grain may be dry enough to transport without needing a grain dryer to further dry it out. In other cases, the moisture content may be so low that simply allowing it to sit and air dry for a certain period of time will be sufficient enough to make it dry grain. In the end, the use of a dryer depends greatly upon the circumstances.

Grain Drying Fundamentals

Drying starts at the bottom of the bin, which is the first place air contacts. The dry air is brought up by the fan through a layer of wet grain. Drying happens in a layer of 1 to 2 feet thick, which is called the drying zone. The drying zone moves from the bottom of the bin to the top, and when it reaches the highest layer, the grain is dry. The grain below drying zone is in equilibrium moisture content with drying air, which means it is safe for storage; while the grain above still needs drying. The air is then forced out the bin through exhaust vent.

Allowable Storage Time

Allowable storage time is an estimate of how long the grain needs to be dried before spoilage and maintain grain quality during storage. In grain storage process, fungi or molds are the primary concern. Many other factors, such as insects, rodents, and bacteria, also affect the condition of storage. The lower the grain temperature is, the longer the allowable storage time will be.

Proper Moisture Levels for Safe Storage

It is possible for long period safe storage if grain moisture content is less than 14%, and stored away from insects, rodents and birds. The following figure is the recommended moisture content for safe storage:

Storage duration	Required MC for safe storage	Potential problems
Weeks to a few months storage	14% or less	Molds, discoloration, respiration loss, insect damage, moisture adsorption
Storage for 8 to 12 months	13% or less	Insect damage
Storage of farmer's seeds	12% or less	Loss of germination
Storage for more than 1 year	9% or less	Loss of germination

Equilibrium Moisture Content

Moisture content in grain is related to the relative humidity and the temperature of the surrounding air. Equilibrium moisture content point is the point when grain no longer losing or gaining water when contacting with drying air. The final moisture content of the grain is up to the amount of moisture in the drying air, which is the relative humidity. The low relative humidity means air is dry and it has a large potential of picking up water. The lower the relative humidity is, the drier the air is. In general, one-half reduce in relative humidity is caused by 20 °C increase in air temperature.

Temperature

Heated air may be used in grain drying process. It can not only accelerate moisture migration inside the kernel, but also can evaporate the moisture on the surface. The major problem about heated air for drying process is the kernel temperature. The grain kernel may be damaged by high kernel temperatures. Usually, kernel temperature is lower than the air temperature. For different use of the corn, temperatures vary. For example, for seed corn, the maximum temperature is 110° F; for livestock feeding corn, the maximum temperature is 180° F.

Aeration

Aeration process refers to the process of moving air through grain. Airflow is a measurement of the amount of air in cubic feet per minute (CFM). In grain drying process, drying time is largely depended on aeration rates. Without sufficient airflow, grain may be damaged before drying is complete. Fans are used to move air through grain.

Application	Grain aeration rate (cfm/bu)
Quality maintenance	1/50 to 1
Natural-air bin drying	1 to 3
Heated-air bin drying	2 to 12
Batch or continuous-flow column dryers	50 to 150
Fluidization	~400

Drying Cost

The drying cost is made up of two parts: the capital cost and the operating cost. Capital cost is largely depend on the drying rate requirement, and equipment cost. Operating cost refers to fuel, electricity and labor force cost. The amount of energy required to dry a bushel of grain is similar for all the drying methods. Some methods depend largely on natural air, while others may use LP heat or natural gas, which make energy cost vary. Basically, fuel and electrical power are the major portions of the operating cost. Drying cost is based on the B.T.U. consumption of temperature change from environment to desired one.

Classification of Grain Drying Methods

In-storage drying methods

- Low-temperature drying
- Multiple-layer drying

Batch drying methods

- Bin batch drying
- Column batch drying

Continuous flow drying methods

- Cross flow drying
- Counter flow drying
- Concurrent flow drying

In-storage Drying Methods

Low-temperature Drying

In-storage drying methods refers to those grain is dried and stored in the same container. Low-temperature drying, also known as near-ambient drying, is one of in-storage drying methods. There are four major factors which influence low temperature drying: the variability of weather, the harvest moisture content, the air flow in the storage bin and the amount of heated air. Most low-temperature dryers are built to dry grain as slowly as possible, while in the same time less spoilage on the grain. It is suggested that low temperature drying system is better operated when the average daily temperature is between 30° C and 50° C. Rather than control the drying air temperature, the low-temperature drying focuses on the relative humidity in order to achieve equilibrium moisture content (EMC) in all grain layers. Low-temperature drying process usually takes 5 days to several months depends on several important variables: weather, airflow, initial moisture content and amount of heat used. Among which, airflow is the key factor. Without appropriate airflow rate, spoilage will occur before drying is completed. By using heated air (LP heat, electric heat and solar heat), the relative humidity of the drying air is better controlled to achieve the desired moisture content. Usually, heated air dryer is used when the relative humidity larger than 70%. In electric heat dryers, an electrical resistant heater is usually placed before the fan to heat the airstream. In some case, a humidistat is employed to control the heater. In solar heat dryers, the drying air passes through the solar collector first to be heated (usually 10 to 12° F rise), then enters the bin through the fan and motor. The advantages of in-storage low temperature drying are quick filling, high quality product, less equipment requirement; while the disadvantages are long drying time, electrical demand if using electric heat, high management skills and uncertain harvest moisture content.

Multiple-layer Drying

Multiple layer drying method refers to the use of LP heat or natural gas in drying corn. Compared to low-temperature methods, multiple-layer drying requires higher temperatures, which results in a shorter allowable storage time. Multiple-layer drying without stirring is the basic multiple-layer drying method, in which airstream is entered through an LP heater by a fan. Usually, the temperature rise after the LP burner is remained low in order to avoid over drying in the bottom layers in the bin. As soon as corn is dried in the bin, the burner is turned off and the fan is used to bring the corn to ambient temperature. The advantages of multiple-layer drying without stirring are little handling of corn, and bin can be used as either dryer or storage; the disadvantages are slow filling and over drying in the bottom layers. Multiple-layer drying with stirring can not only dry grain equilibrium from top to bottom, but also decrease the air resistance of the grain. Moreover, using stirring system can avoid over drying in bottom layer problem and give a uniform grain moisture content in the whole bin. When drying is complete, the burner is turned off while the fan and stirrer are used to mix the corn to achieve equal moisture content and temperature. The advantages of adding stirring are preventing over drying and accelerating drying and allowable fill rate; the disadvantages of stirring system are additional expenses and decreasing bin capacity.

Batch Drying Methods

Bin-batch Drying

In batch drying methods, certain amount of grain is placed first, usually 2 to 4 inches, the batch is dried and cooled later, then drying is stopped and batch is removed. The batch dryers are usually operating under this sequence and repeating this sequence for several times. The bin-batch drying methods employ a full perforated floor as the dryer. Without stirring, large variety of equipment is available and the batch can be used as both dryer and cooler, but there may be large moisture gradient from top to bottom and losing time in loading and unloading process. When adding stirring system, unequilibrium moisture content problem is avoided, however, stirrer is an added expense. When using bin-batch roof dryer, time losing problem can be solved. There is a drying floor under the bin roof and the drying fan and burner is installed high on the bin wall. When the drying process is completed, grain is put in the regular bin floor, thus unloading time is reduced. However, there is no wet grain holding in bin-batch roof dryers and there is more expense on machines.

Column Batch Drying

The column formed in this kind of dryer is made up of two vertical perforated steel sheets, which is about 12 inches thick each. The capacity of column batch dryers is too small to store grains. The advantages of column batch, stationary bed dryer are easy to move and the dyer can be used as cooler; while the disadvantages are time losing when cooling, loading and unloading and unequal moisture distribution when drying is completed. When column batch recirculating dryer is used, the moisture content variation problem is avoided, but the additional handling process may result in grain spoilage.

Continuous Flow Drying Methods

Cross Flow Drying

Cross flow dryers is one of the most widely used continuous flow dryers. In the cross flow dryer, the airstream is perpendicular to the grain flow. Then the grain near the drying air is over dried, while on the other side, grain is under dried. Moisture gradient exists when drying is complete. In reality, the lower the airflow rate, the higher the grain moisture content variation between two sides of the column.

Concurrent Flow Drying

In the concurrent flow dryer, both the grain and air are moving in the same direction, which means the wettest grain is subjected to the hottest drying air. The kernels leave the drying region at the same temperature and the same moisture content. Energy efficiency is 40% better compared to cross flow dryer. However, the bed depth must be deeper than 12 inches depth than cross flow type. Thus, fan power requirements are high in this type of dryer.

Counter Flow Drying

In the counter flow dryer, the grain and the air are moving in opposite directions, which mean the driest grain is subjected to the hottest drying air. The kernels leave the drying region at the same temperature and the same moisture as in concurrent flow dryers. The suggesting air temperatures are less than 180° F because the driest kernels are more likely to be damaged by hot air.

Applications of Grain Drying

Sunflower Drying

For different types of sunflowers, the preservation moisture content is different. Oilseed sunflowers are better dried to 9 percent moisture content, while bird seed sunflowers are 10 percent moisture content. Compared to corn drying, sunflowers are more easily dried and kept in safe storage.

What's more, high temperature may not have adverse effect on sunflowers kernel, which may be the reason of the fatty acid composition. There was no evidence of damage when air was heated up to 220° F in drying. However, fine hairs and fibers on the seed coat of sunflowers may cause fire hazard. It is suggested that remove the flaming particles first when heating the sunflowers.

Bean Drying

The seed coat of bean is quite fragile and easy to damaged in cracking and splitting, which may cause loss to the producer. Some studies on beans suggested that in order to avoid cracking, it is better to keep drying air above 40 percent relative humidity.

Corn Drying

When drying corn kernels it is important to keep in mind that cracks and fractures in the corn can lead to many problems in both the storage and processing. The major problem that occurs from high temperature drying and then rapid cooling of the grain is stress-cracking. Stress-cracking is when fractures become present in the corn endosperm. Stress-cracked kernels often absorb water too quickly, are more likely to become broken, and are increasingly susceptible to insect and mold damage during dry storage. In order to reduce the amount of grain that is lost due to stress-cracking, medium temperature and slow cooling, or natural air and low temperature drying methods should be employed.

Rice Huller

A rice huller is a machine used in rice cultivation for removing the chaff from the grains of rice. There have been many innovations and techniques to hull rice. Traditionally it was pounded using a mortar and pestle. The machine called the rice pounder was an early invention. Various such innovations were developed throughout Asia and were constantly upgraded. The Engelberg huller is a more modern innovation, developed by an engineer in Brazil and patented in 1885.

The Engelberg huller has steel rollers for removing the husk. Some other variants of this rice huller may have disk or cono hullers, which uses an abrasive rotating disk for removing the husk before the grain is passed to the conical rollers, which polish it. The process is done repeatedly until all sides of the rice grains are husked. To reduce grain breakage, rubber rollers are used. The only drawback to this is that these rollers require frequent replacement.

Bale Lifter

A bale lifter is a farm equipment used to transport hayor straw bales. It is connected to a tractor via the three-point linkage. They come in either double or single bale lifters and are usually fitted with a tipping device to tip the round bales on to their flat ends so that the bales do not deform when lying over the winter.

Hay Rake

A hay rake is an implement used in agriculture that collects the straw or cut hay into windrows for collection later. It is designed such that it can be used for fluffing up the hay and turning it over for drying. It can also be used to protect the hay from morning dew. On the next morning, the hay is spread again with a tedder so that it dries quickly. A hay rake can be drawn by animals or by a tractor. It can also be used like a hand tool. The earliest designs were mostly wooden hand rakes having wooden teeth, which still are prevalent in some parts of the world.

An early variant, the horse-drawn hay rake had iron teeth and was operated from a seat that was mounted over the rake and equipped with a lever-operated lifting mechanism. This rake was used to gather cut hay into windrows. A sweep rake that is either horse-drawn or tractor-mounted, could be used to gather the windrowed hay and then for loading onto a wagon. A later variant called the side delivery rake was developed, which had a chain-driven or a gear-driven reel mounted approximately at an angle of 45 degrees, so that the hay was gathered and pushed to one side of the rake as it was drawn across the field. The side delivery rake was convenient because it eliminated the processes of back-and-forth raking, raising and lowering which are typical operations using a dump rake. Much innovation was done to this rake with a higher frame system and severe transverse angles.

A variety of wheel rakes were developed which employed different numbers of spring-tooth encircled wheels that were mounted on a frame and driven by free-wheeling contact with the forward motion of the implement. Such rakes were popularized for their simpler mechanics, low maintenance, cheap operability and for its gentle action on the hay.

Permissions

All chapters in this book are published with permission under the Creative Commons Attribution Share Alike License or equivalent. Every chapter published in this book has been scrutinized by our experts. Their significance has been extensively debated. The topics covered herein carry significant information for a comprehensive understanding. They may even be implemented as practical applications or may be referred to as a beginning point for further studies.

We would like to thank the editorial team for lending their expertise to make the book truly unique. They have played a crucial role in the development of this book. Without their invaluable contributions this book wouldn't have been possible. They have made vital efforts to compile up to date information on the varied aspects of this subject to make this book a valuable addition to the collection of many professionals and students.

This book was conceptualized with the vision of imparting up-to-date and integrated information in this field. To ensure the same, a matchless editorial board was set up. Every individual on the board went through rigorous rounds of assessment to prove their worth. After which they invested a large part of their time researching and compiling the most relevant data for our readers.

The editorial board has been involved in producing this book since its inception. They have spent rigorous hours researching and exploring the diverse topics which have resulted in the successful publishing of this book. They have passed on their knowledge of decades through this book. To expedite this challenging task, the publisher supported the team at every step. A small team of assistant editors was also appointed to further simplify the editing procedure and attain best results for the readers.

Apart from the editorial board, the designing team has also invested a significant amount of their time in understanding the subject and creating the most relevant covers. They scrutinized every image to scout for the most suitable representation of the subject and create an appropriate cover for the book.

The publishing team has been an ardent support to the editorial, designing and production team. Their endless efforts to recruit the best for this project, has resulted in the accomplishment of this book. They are a veteran in the field of academics and their pool of knowledge is as vast as their experience in printing. Their expertise and guidance has proved useful at every step. Their uncompromising quality standards have made this book an exceptional effort. Their encouragement from time to time has been an inspiration for everyone.

The publisher and the editorial board hope that this book will prove to be a valuable piece of knowledge for students, practitioners and scholars across the globe.

Index